The Urban Book Series

The Urban Book Series is a resource for urban studies and geography research worldwide. It provides a unique and innovative resource for the latest developments in the field, nurturing a comprehensive and encompassing publication venue for urban studies, urban geography, planning and regional development.

The series publishes peer-reviewed volumes related to urbanization, sustainability, urban environments, sustainable urbanism, governance, globalization, urban and sustainable development, spatial and area studies, urban management, transport systems, urban infrastructure, urban dynamics, green cities and urban landscapes. It also invites research which documents urbanization processes and urban dynamics on a national, regional and local level, welcoming case studies, as well as comparative and applied research.

The series will appeal to urbanists, geographers, planners, engineers, architects, policy makers, and to all of those interested in a wide-ranging overview of contemporary urban studies and innovations in the field. It accepts monographs, edited volumes and textbooks.

Indexed by Scopus.

Grazia Brunetta · Patrizia Lombardi ·
Angioletta Voghera
Editors

Post Un-Lock

From Territorial Vulnerabilities to Local Resilience

Politecnico di Torino · Università di Torino · Interuniversity Department of Regional and Urban Studies and Planning, Excellence MIUR 2018-2022 · R3C Responsible Risk Resilience Centre · SDG11 lab · Springer

Editors
Grazia Brunetta
DIST—Interuniversity Department of Regional and Urban Studies and Planning
R3C—Interdepartmental Responsible Risk Resilience Centre
Politecnico di Torino
Turin, Italy

Patrizia Lombardi
DIST—Interuniversity Department of Regional and Urban Studies and Planning
Politecnico di Torino
Turin, Italy

Angioletta Voghera
DIST—Interuniversity Department of Regional and Urban Studies and Planning
R3C—Interdepartmental Responsible Risk Resilience Centre
Politecnico di Torino
Turin, Italy

ISSN 2365-757X ISSN 2365-7588 (electronic)
The Urban Book Series
ISBN 978-3-031-33893-9 ISBN 978-3-031-33894-6 (eBook)
https://doi.org/10.1007/978-3-031-33894-6

This Springer imprint is published by the registered company Springer Nature Switzerland AG
The registered company address is: Gewerbestrasse 11, 6330 Cham, Switzerland

Preface

This book is the first result of the "Post Un-Lock" research project, funded by the Interuniversity Department of Regional and Urban Studies and Planning (DIST) of the Politecnico di Torino, a three-year work involving the collaboration of a large number of researchers belonging not only to different departments and institutions but also to diverse disciplinary areas.

The research involves teams from the Department of Environmental, Land, and Infrastructure Engineering (DIATI) of the Politecnico di Torino (such as environmental and geomatics engineers), the Department of Mathematical Sciences "Giuseppe Luigi Lagrange" (experts in mathematical statistics), and the Department of Medical Sciences of the Università di Torino (researchers in the field of medical statistics and epidemiology), committed to reflecting on the territorialization of the pandemic and the effects of pollution on health in cities.

The Post Un-Lock research took shape in the last months of 2020, triggered by the scientific impulse of the international goals of sustainability and resilience in the perspective of the post-carbon city and the overcoming of environmental, economic, social and health crises. Scholars have collaborated and tested on case studies methodologies, approaches, and tools capable of re-imagining cities and regions to overcome vulnerabilities and to innovate the socio-ecological system on the basis of the ideal-typical model of Local Resilience Unit.

With this in mind, we would like to thank, in addition to the authors of the chapters and their collaborators: the partners of the research such as the Interdepartmental Centre Responsible Risk Resilience Centre (R3C), the SDG11LAB, the S3+Lab (Urban Sustainability and Security Laboratory for Social Challenges), the CED PPN (European Documentation Center on Natural Park Planning), the Living Lab and the PIC4SER of Politecnico di Torino. We would also like to thank the external supporting institutions such as IUGA of Grenoble, University of South Denmark, CMCC (Euro-Mediterranean Center on Climate Change), and the administrations involved in the case studies, whose suggestions provided crucial support for the outcomes of this publication. Other institutions and people we would like to thank are: Nicola Tollin, Professor with special responsibilities in Urban Resilience, UNESCO Chair at University of Southern Denmark (SDU), ITI,

Civil and Architectural Engineering; Joe Ravetz, Co-director of the Collaboratory for Urban Resilience and Energy at the Manchester Urban Institute, University of Manchester; Jordi Morató, Director UNESCO Chair on Sustainability at Polytechnical University of Catalonia; Michele Talia, Istituto Nazionale di Urbanistica (INU); Stefania Crotta, Direzione Ambiente, Energia e Territorio of Regione Piemonte; Milena Maule, Associate Professor of Medical Statistics and Epidemiology at Department of Medical Sciences of Università di Torino (UNITO); Egidio Dansero, Professor at Department of Cultures, Politics, and Society (UNITO) and delegate at RUS; Alessio Malcevschi, Professor at Università di Parma, RUS delegate; Stefano Armenia, Senior Research Fellow at Link Campus University, president of System Dynamics—SYDIC; Rosa Gilardi, Director of Area Urbanistica e Qualità dell'Ambiente Costruito, Città di Torino; Gilles Novarina, Laboratoire Architecture Environnement and Cultures Constructives at Université Grenoble Alpes—École Nationale Supérieure d'Architecture; Guglielmo Filippini, Direzione Risorse Idriche e Tutela dell'Atmosfera at Ufficio Pianificazione e Controllo delle Risorse Idriche of Città Metropolitana di Torino.

Turin, Italy
January 2022

Grazia Brunetta
Patrizia Lombardi
Angioletta Voghera

Contents

Part I
The Research

Chapter 1
Post-pandemic Challenges. The Role of Local Governance for Territorial Resilience

Grazia Brunetta and Angioletta Voghera

Abstract This chapter aims to provide researchers different interpretative keys of the book, which attempt to propose methodologies, tools, and case studies to put resilience into action in post-pandemic territories by planning and design at different scales. The chapter opens the discussion by presenting diverse and interdisciplinary contributions of which the research is composed; it discusses key topics with reference to the transformative resilience, and referring to methodologies and tools for interpreting territories, and focusing on the role of planning, as well as attempting to describe through practices the operational concept of the Local Resilience Unit.

Keyword Territorial resilience · Local resilience unit · Methodologies · Case studies

1.1 Framing Territorial Resilience

What is the role of urban and regional planning in achieving sustainable development goals of our communities, considering the turning point for spatial planning posed by the COVID-19 emergency? Moreover, how do medium and long-term planning purposes interact with the needs that the pandemic has boosted?

The book *Post Un-Lock—From territorial vulnerabilities to local resilience* aims to provide the reader with a valuable tool to understand how the COVID-19 pandemic should be a trigger for the re-start based on a new territorial path in the frame of sustainability and resilience.

As shared among scholars, sustainability and resilience are two related visions. They are two umbrella issues that cannot be used as interchangeable notions, since

G. Brunetta · A. Voghera (✉)
DIST—Interuniversity Department of Regional and Urban Studies and Planning,
R3C—Interdepartmental Responsible Risk Resilience Centre, Politecnico di Torino, Torino, Italy
e-mail: angioletta.voghera@polito.it

G. Brunetta
e-mail: grazia.brunetta@polito.it

G. Brunetta et al. (eds.), *Post Un-Lock*, The Urban Book Series,
https://doi.org/10.1007/978-3-031-33894-6_1

resilience can be considered a subsystem of the sustainability approach (Folke et al. 2016; Brunetta and Caldarice 2020) or a renewed and systemic attitude in the sustainability science (Xu et al. 2015). Sustainability and resilience have some differences (Voghera and Giudice 2019): sustainability promotes general aims of social justice, environmental protection, and economic efficiency through a radical re-organization of the socio-ecological system; resilience is focused on the need for change, adaptation, and transformation of territorial systems, evolving in time and space following the socio-ecological demands and overcoming the socio-ecological limits. Besides, they have shared features, e.g., supporting the integration of environmental management into territorial and urban planning, reinforcing reflectiveness, flexibility, and inclusiveness in planning processes, and valorizing the robustness and diversity of ecological and landscape values.

Territorial innovation is required to guarantee sustainability, resilience, and the related ecological transition. This pathway implies the "bounce forward" of territorial systems, demanding them to evolve in a post-COVID process of change. This bounce calls for the re-design of new planning solutions. Moreover, it is necessary to renew the current tools, as well as to adopt new approaches, including technological ones, an envisage actions and policies for the resilience of territories (Voghera 2020). Assuming a strengthened definition, the IPCC (2012) frames resilience as "*the ability of a system and its component to anticipate, absorb, accommodate or recover from the effects of a hazards in a timely and efficient way, including by ensuring the preservation, restoration or enhancement of its essential basic structures and functions.*" In this rationale, the territorial dimension is an integrated layer, and "territorial resilience" can be described as "*an emerging concept (…) that considers the dynamic and nonlinear interaction between endogenous features of systems and their exogenous transient shocks, chronic stresses, and their conditions of sudden or slow change that affect both assets and governance systems*" (Brunetta et al. 2019, p. 10). This definition emphasizes the territorial ability to respond, transform, and co-evolve into a different perspective without rebounding back after the pandemic event (Giovannini et al. 2020). This latter definition, the so-called transformative resilience, should take advantage of the environmental, economic, climate, and social crises, and reconsider planning and design processes.

The Post Un-Lock research supports a jump forward on "territorial resilience" based on interpreting the "inherent unpredictability" of the pandemic and post-pandemic times (Folke et al. 2021, p. 19). This complex leap ahead should enable to reorganize of territorial systems, starting from the awareness of interrelation among all their components—such as economic, social, ecological, energy, climate, and health—and following the directions drafted by the UN Agenda 2030 (Agenda for Sustainable Development, 2015) and by the different European Next Generation EU Plans (in Italy the PNRR).

Spatial policies assume a leading role in sustainable and resilient development. They have not only considered system dynamics but also encouraged our collective capacity to guide development trajectories in a responsive, adaptive, and reflective way (Freeman et al. 2020). What is needed is a process that moves towards nature-based recovery or towards a nature-positive economy (IUCN Marseille Congress

2021 "Our Nature, Our Future"). This process should be based on the integration of site-based conservation and development (see Chap. 3) as a path to rewriting the alliance between humans and nature, reversing the paradigms of the past 100 years (today, 0.01% of living things have been able to produce materials/objects/waste equal to the mass of all living things; "Great Acceleration"; Elhacham et Al. 2020).

Considering territorial governance to implement SDGs 3, 11, 13, 14, and 15 required a significant acceleration by integrating the ecological transition into planning and management at different scales (from the global consciousness to the local action).

Radical changes in spatial planning at different scales are the basis of transformative resilience. The demand for innovation has been discussed for thirty years in spatial planning. Otherwise, planning has produced partial, sectorial, and fragmented responses to the demands for innovation. Planning has attempted to pay attention to many issues, such as land consumption, energy conversion of the built environment, ecological and landscape design, social services, and transportation. These attentions have not yet produced different organizations of territories concerning the environment and landscape to ensure health, safety, and well-being.

With this in mind, the research tried to answer the following emerging questions:

- We are certainly on a proactive ground, but do we have appropriate cognitive and planning tools for the complex challenges and goals for ecological transition and resilience?
- What is the role of planning in driving these goals to action?
- Can planning lead to effective ecological transition actions?

In a nutshell, the COVID-19 experience highlights the need to suggest a planning system that can integrate multiple scales according to an interdisciplinary approach focused on deep knowledge of territorial risks and vulnerabilities.

1.2 Territorial Resilience. Does the Scale Matter?

The operationalization of spatial resilience at different planning scales depends on local conditions and specific spatial policy objectives. Within the framework of international agreements, it is now clear that territorial resilience, to be effective, must reinforce a multiscalar governance approach. Vertical integration should address how local action contributes to international and national policy goals and how national governments can sustain local action (Caldarice et al., 2021).

From a practical point of view, the EU Green Deal (2019) represents the first attempt to decline resilient thinking into public policy, promoting multiscalar policies to develop a climate-neutral development model by 2050 both at the regional and at the local scales.

At the regional scale, territorial resilience should be the frame of a strategic vision that fields a network of policies based on shared agreements between stakeholders and institutions, capable of supporting the empowerment and accountability of each actor, guaranteeing the adequacy of regional planning tools, and identifying economic resources to integrate the climate transition to resilience into decision-making processes (Brunetta 2015).

At the local scale, territorial resilience should be the lead for the territorial regulatory framework to support design actions that sustain a green and equitable transition. The current situation requires new paradigms and approaches for territorial governance design, aiming to link measures to reduce territorial vulnerabilities with strategies and interventions to guide territories' transformation, maintenance, and territorial regeneration design. Territorial resilience requires two equally considered dimensions, the strategic and the local, which must work together in circular and mutual processes (Coscia and Voghera 2022, 2023). These two " drivers" of territorial resilience imply innovative cognitive frameworks in decision-making, from analysis of territorial vulnerabilities to planning actions.

During the COVID-19 emergency, the "local resilience unit" emerged as a resilient answer at the local scale, emphasizing the role of community proximity and the liveability of urban spaces and facilities. In this book, the local resilience unit—that is a specific outcome of the research—can be defined as "an operational frame" at the "neighbourhood" level that can develop planning actions together with community empowerment to make cities more responsive, resilient, and able to provide a high level of liveability and urban well-being. Of course, this is not a completely new planning paradigm. The Local Resilience Unit stems from an ideal–typical model of territorial organization based on civic subsidiarity that can provide an operational key to integrate local demands, local self-organization, and the responses of public institutions (Brunetta and Moroni 2012).

Building on this theoretical overview, the book brings together several contributions that address various open issues, e.g., understanding spatial, landscape, environmental, and climate dynamics, analysing local vulnerabilities, and using modern survey techniques and tools to produce planning support. By proposing the Local Resilience Unit, Post Un-Lock is taking a step towards defining a new spatial planning paradigm that deals with territorial transformative resilience, aiming at supporting the implementation of innovative practices and actions for overcoming territorial vulnerabilities.

1.3 Converging Experimentations: Challenges, Methodologies, and Tools for Post-pandemic Territories and Cities

The book presents the results of work carried out by researchers at the Polytechnic and University of Turin who collaborated in the "*POST-UN-LOCK. From territorial vulnerabilities to local resilience*" research program.[1] The project's main objective is to decline through case study a first conceptual definition of "local resilience units," and, more broadly, of planning for resilience in the post-pandemic period. In this perspective, there is a strong focus on the definition of a new paradigm focused on the importance of local and "sub-local" planning, considered as the output of a deep relationship between territories and the knowledge paradigm. The concept of Local Resilience Unit is a recent topic that is still open in the literature. It is intended as a "micro-territory" capable of responding to shocks, the Local Resilience Unit is linked to the concept of neighbourhood unit, superblock, and 15 Minute city. It should be understood not only as a form of optimal distribution of essential services, organized and planned for enabling communities to overcome crisis through an adaptive approach, reinventing the territorial proximity.

The aim is to provide, through the concept of a resilience unit and an integrated reading of risks and vulnerabilities, an operational framework for post-COVID planning. The objective is to provide, on the one hand, ideas for an integrated interpretation of territorial risks and vulnerabilities and, on the other, practical and theoretical models for the post-COVID city, capable of reorganizing itself to pursue long-term sustainability and resilience. In this process of definition, it is necessary to work at different scales, considering how a specific system of risks to which a resilience unit is subjected is the result of the interaction of local and supra-local phenomena. Moreover, the response to these risks involves site-specific actions that require a high level of territorial awareness at different scales and the integration of policies among different levels of government and between different spheres according to a multiscalar and transdisciplinary approach.

The research POST UN-LOCK provides a wide range of reflections dealing with methodologies, approaches, and tools experimented on different territories: the Piedmont Region, the Metropolitan City of Turin, the Stura River, and the Lanzo Valley territory.

The work of the researchers in the book was organized into three sections: topics, case studies, and digital tools.

The section **Topics** tries to define some main issues for the resilient city critically interpreted through diverse qualitative and quantitative methodologies used to analyse the territory. It explores some specific aspects:

- the spatial analysis of the COVID-19 pandemic and its effect on Piedmont municipalities (Chap. 2);

[1] The research POST-UNLOCK. From territorial vulnerabilities to local resilience was financed by DIST Department in 2020.

- an interpretation of hydrological, ecological, and ecosystem features of the territory (Chap. 3);
- the ecological network and ecosystem services for territorial resilience (Chap. 4);
- an evaluation of sustainability at the local level (Chap. 5);
- an interpretation of the concept of neighbourhood concerning the minimum resilience units (Chap. 6).

The section **Case Studies** develops reflections on various territorial scale (from the metropolitan to the neighbourhood scale) reflecting on specific aspects, such as NO2 concentrations and COVID-19 in Turin Municipality (Chap. 7); the pandemic event in mountain areas compared to the metropolitan area (Chap. 8); risks and climate change in North Turin metropolitan area (Chap. 9); experimentations for landscape resilience at local level (Chap. 10); the school and social innovation for territorial resilience and Local Resilience Unit (Chap. 11).

The **Digital Tools section** describes experimentations that are essential to developing the resilient challenges and investigating specific interpretation tools, considering 3D metric surveys for the digital cartographic production (Chap. 12), and the role of sources and data for the analysis used by GIS tools (Chap. 13).

Some finals remarks and perspectives are reported in Chap. 14.

The analysis and studies of the territory, which are also conducted with the help of GIS methodologies (in the three parts of the book topic, case studies, and digital tools), are essential for the knowledge on both the large and local scales of local risks and vulnerabilities. Besides, those tools have a crucial role in constructing theoretical (and critical) interpretations of the main problems and solutions that can converge in the "local resilience unit" concept.

References

Brunetta G (ed) (2015) Smart evaluation and integrated design in regional development. Territorial scenarios in Trentino, Italy. Ashgate, Farnham

Brunetta G, Caldarice O (2020) Spatial resilience in planning: meanings, challenges, and perspectives for urban transition. In: Leal Filho W, Azul A, Brandli L, Özuyar P, Wall T (eds) Sustainable cities and communities. Encyclopedia of the unsustainable development goals, vol III. Springer International Publishing, Cham, pp 1–12. ISBN: 9783319957166. https://doi.org/10.1007/978-3-319-95717-3_28

Brunetta G, Moroni S (2012) Contractual community in the self-organising City. Freedom, Creativity, Subsidiarity. Springer, Dordrecht

Brunetta G, Ceravolo R, Barbieri CA, Borghini A, de Carlo F, Mela A, Beltramo S, Longhi A, De Lucia G, Ferraris S, Pezzoli A, Quagliolo C, Salata S, Voghera A (2019) Territorial resilience: toward a proactive meaning for spatial planning. Sustainability 11(8):2286, 17, ISSN 2071-1050. https://doi.org/10.3390/su11082286. www.mdpi.com/journal/sustainability

Caldarice O, Tollin N, Pizzorni M (2021) The relevance of science-policy-practice dialogue. Exploring the urban climate resilience governance in Italy. City, Territory Archit 8:1–11, ISSN: 2195-2701. https://doi.org/10.1186/s40410-021-00137-y

Coscia C, Voghera A (2022) Co-evolutionary, transformative, and economic resilience during the COVID-19 pandemic crisis. Evidence-based experiences of urban community design in Turin

(Italy). In: Calabrò F, Della Spina L, Piñeira Mantiñán MJ (eds) Lecture notes in networks and systems. new metropolitan perspectives—post covid dynamics: green and digital transition, between metropolitan and return to villages perspectives, vol 2. Springer, Cham, Berlino. https://doi.org/10.1007/978-3-031-06825-6_271

Coscia C, Voghera A (2023) Resilience in action: The bottom up! architecture festival in Turin (Italy). J Safety Sci Resilience 4(2):174–187. https://doi.org/10.1016/j.jnlssr.2022.10.005

Elhacham E, Ben-Uri L, Grozovski J et al (2020) Global human-made mass exceeds all living biomass. Nature 588:442–444. https://doi.org/10.1038/s41586-020-3010-5

Folke C, Biggs R, Norström AV, Reyers B, Rockström J (2016) Social-ecological resilience and biosphere-based sustainability science. Ecol Soc 21(3):41. https://doi.org/10.5751/ES-08748-210341

Folke C, Polasky S, Rockström J et al (2021) Our future in the Anthropocene biosphere. Ambio 50:834–869. https://doi.org/10.1007/s13280-021-01544-8

Freeman J, Baggio JA, Coyle TR (2020) Social and general intelligence improves collective action in a common pool resource systems. Proc Natl Acad Sci 117(14)

Giovannini E, Benczur P, Campolongo F, Cariboni J, Manca A (2020) Time for transformative resilience: the COVID-19 emergency. Publications Office of the European Union, Luxembourg. https://doi.org/10.2760/062495

IPCC (2012) Glossary of terms. In: Managing the risks of extreme events and disasters to advance climate change adaptation. In: Field CB, Barros V, Stocker TF, Qin D, Dokken DJ, Ebi KL, Mastrandrea MD, Mach KJ, Plattner G-K, Allen SK, Tignor M, Midgley PM (eds) A special report of working groups I and II of the intergovernmental panel on climate change (IPCC). Cambridge University Press, Cambridge, UK, and New York, NY, USA, pp 555–564

Voghera A, Giudice B (2019) Evaluating and planning green infrastructure: a strategic perspective for sustainability and resilience. Sustainability 11(10):2726, 21, ISSN 2071-1050. https://doi.org/10.3390/su11102726. https://www.mdpi.com/2071-1050/11/10/2726/htm

Voghera A (2020) The river agreement in Italy. Resilient planning for the co-evolution of communities and landscapes. Land Use Policy 91. Press, available online 30 November 2019, published February 2020, ISSN: 0264-8377, https://doi.org/10.1016/j.landusepol.2019.104377

Xu L, Marinova D, Guo X (2015) Resilience thinking: a renewed system approach for sustainability science. Sustain Sci 10:123–138. https://doi.org/10.1007/s11625-014-0274-4

Part II
Topics

Chapter 2
Notes on Spatial Implications of COVID-19. Evidence from Piedmont Region, Italy

Grazia Brunetta, Ombretta Caldarice, Danial Mohabat Doost, and Franco Pellerey

Abstract The COVID-19 pandemic has been studied by many scholars from a wide range of disciplines. Among these investigations, planners and regional scientists have researched the spatial spread of the contagion. Most of these studies tried to explore the spread of the disease in a fixed period, like annually, and analysed the spatial variables that are most influential on the COVID-19 spread over territories. On the same line, the chapter investigates the pattern of virus transmission in the 1.181 municipalities of the Piedmont Region during two years of the pandemic over different periods by providing 24 monthly and two annual hot spot maps using the Spatial Statistics Tools on ArcGIS. Consequently, the chapter analyses the correlation between the spread of contagion with three spatial variables (population density, annual average traffic flow, and the ageing index) by performing a statistical analysis on the municipalities which showed unexpectedly higher or unexpectedly lower numbers of contagion. The results show that the impacts of population density and annual average traffic flow are verified on the transmission rate of the cities with unexpectedly higher or lower exposure to COVID-19 contagion than their neighbours both in the first and the second year of the pandemic. For the ageing index, an association is noticed during the first year while not confirmed for the second. In conclusion, the chapter proposes that studying the disease's variations—at different times and on a regional scale—uncovers the spatial dimension of the phenomenon and would suggest insights for both scientists and policymakers to enrich preparedness as the preferable approach in future planning policies towards transformative resilience.

Keywords COVID-19 pandemic · Transformative resilience · Piedmont region · Spatial planning · Geographic information system

G. Brunetta · O. Caldarice (✉) · D. Mohabat Doost
DIST—Interuniversity Department of Regional and Urban Studies and Planning, R3C—Interdepartmental Responsible Risk Resilience Centre, Politecnico di Torino, Torino, Italy
e-mail: ombretta.caldarice@polito.it

F. Pellerey
DISMA—Department of Mathematical Sciences "Giuseppe Luigi Lagrange", R3C—Interdepartmental Responsible Risk Resilience Centre, Politecnico di Torino, Torino, Italy

G. Brunetta et al. (eds.), *Post Un-Lock*, The Urban Book Series,
https://doi.org/10.1007/978-3-031-33894-6_2

2.1 How Do Spatial Implications Matter in the COVID-19 Pandemic?

In his 2012 seminal book, David Quammen (2012) defines pandemics as *spillovers*, questioning if they are merely happening to us or if we are somehow causing them. Most recently, Nassim Taleb reminded us that the COVID-19 pandemic is not a *black swan* but a warning of a more fragile global system (Avishai 2020). Precisely for this reason, the ongoing COVID-19 crisis powerfully reveals its spatial implication (Coppola 2021). On the one hand, spatial relations can intensively contribute to the diffusion of the disease; on the other hand, this contribution might change over time during the pandemic, so both temporal and spatial dimensions are essential aspects to be studied.

Along this line, planners and regional scientists are currently engaged in addressing the COVID-19 physical dimension, as it is a phenomenon that establishes spatial features in addition to the human one (Allain-Dupré et al. 2020). Although the health effects of COVID-19 are not neglectable, the impacts on the population go beyond the spread of the disease.

In addition to the temporal dimension—that only a few recent kinds of research considered for analysing the pandemic's territorial changes—it is also helpful to find the most influential spatial variables that might contribute to the spread of infection. Despite the growing awareness about the spatial implication of COVID-19, a limited number of place-based studies have been published since the initial pandemic outbreak. Among the existing literature, some scholars traditionally relate density and crowding to the intensification of the disease spread (Wong and Li 2020), while a growing number of researchers suggest either weak, strong or even no correlation between density and COVID-19 spread (Carozzi et al. 2020). Many empirical researchers perform case study analyses at the national level to identify the role of density as a predictor of infection rates. Among those, Bhadra et al. (2021) suggest a moderate association between COVID-19 spread and density in India, while Kadi and Khelfaoui (2020) demonstrate a stronger correlation between the two in the case of Algeria. In addition to density, some other studies have attempted to reveal spatial parameters which can explain the diffusion of COVID-19. These factors mainly include connectivity and ageing. On this line, Hamidi et al. (2020) claim that connectivity matters more than density in the US spread of COVID-19, while Mollalo and Vahedi (2020) recognize ageing and demographic variables as central.

Turning to the Italian context, the empirical studies of COVID-19 partially concentrate on the spatial dimension of contagion. Buja et al. (2020) showed that the spread of COVID-19 was correlated negatively with the ageing index in provinces of Northern Italy, underlying that provinces with younger populations are more likely to be the hot spots for the rapid spread of the disease. Dettori et al. (2021) compared the pattern of COVID-19-related mortality in Italy with several geographical, environmental, and socioeconomic variables at the provincial level, demonstrating that contagious is related to pollutants and land take.

Starting from this broad framework, this chapter initially performs a territorial analysis on the increase of the pandemic using the GIS-based hot spot analysis regarding the spread of the COVID-19 infection in the 1.181 municipalities of the Piedmont Region by providing 24 monthly and two annual maps to track the movement of the disease outbreaks over time. The first set of results shows that the hot spots and cold spots of contagion are functions of both time and space, and different episodes move within the region. By changing the spatial scale of the analysis, it is observed that some municipalities—even inside the clusters of hot spots and cold spots—show an unexpectedly lower or higher contagion rate compared to their neighbours. Consequently, we analysed the correlation between the spread of contagion with three spatial variables (population density, annual average traffic flow, and the ageing index) by performing a statistical analysis on the municipalities which showed unexpectedly higher or unexpectedly lower numbers of contagion.

From this perspective, the chapter aims to fill the gap as there are no studies performed at the local level in Italy, to the best of our knowledge. Accordingly, the novelty of this study refers to its methodological approach in considering spatial and scalar factors of the pandemic in a way that:

- It is performed through a geostatistical analysis using GIS techniques and considering the proximity factors between municipalities.
- It is not merely limited to the analysis of density as a spatial variable, integrating connectivity and ageing as well.
- It is developed at the local scale, fitting the Italian planning system.

In the end, this study aims to suggest implications for policymakers to determine adequate courses of action to solve the problem or implement reforms independently from the pandemics, enriching preparedness (Balducci 2020) with an adaptation path in a "build back better" approach towards transformative resilience (Brunetta and Caldarice 2020).

2.2 Methodology

As mentioned earlier, this study performed a GIS-based geospatial analysis for 1.181 Piedmont Region municipalities located in northwest Italy with an area of 25,402 km^2, making it the second-largest region of the country, with more than 4 million population as of 31 January 2021. The Piedmont Region has eight provinces, including the metropolitan area of Turin and the city of Turin as the capital. The geographical divisions of the country, regions, provinces, and municipalities of the study area are visible in Fig. 2.1.

In Fig. 2.2, the methodological workflow of this study is presented, which consist in several steps, including data collection, data processing, applying the hot spot test to find out how clusters moved within the territory, performing EBK for finding the cities with unexpectedly higher or lower exposure to contagion, and analysing the

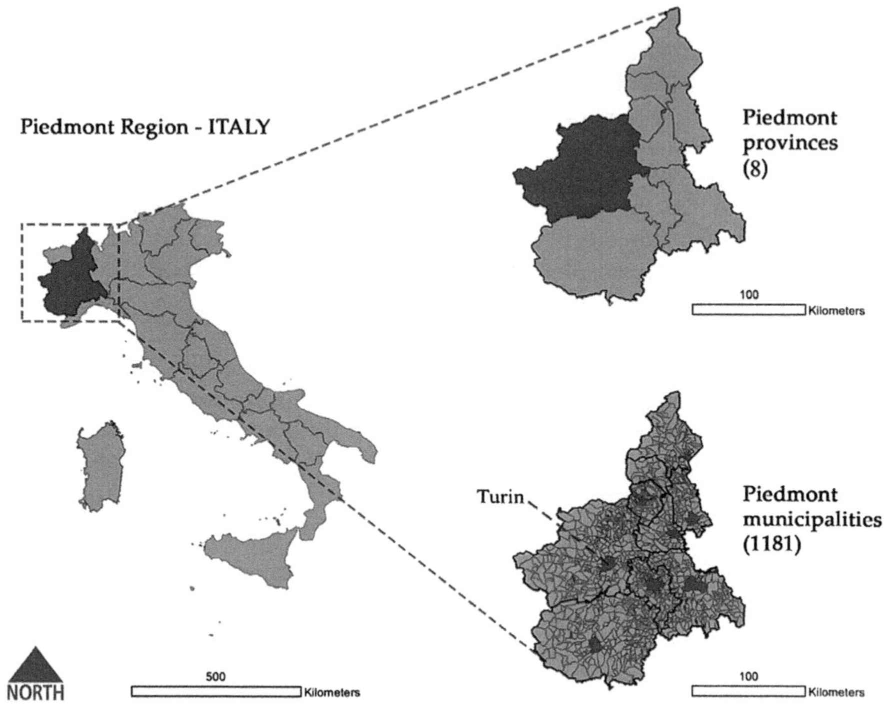

Fig. 2.1 Localization of the study area

territorial variables of these municipalities. All these steps will be explained in detail followingly.

As the first step, the raw data are extracted from the daily updated maps prepared and published every day by the Piedmont Region website. These daily updated maps provide the number of present-day positive cases in every 1.181 municipalities of the Piedmont Region. It is important to note that these numbers represent the new cases for each day besides those who tested positive in the previous days and are still positive. The data have been extracted from 26 March 2020 to the end of February 2022, enabling us to cover two years of the pandemic. In addition, the demographic and territorial data are compiled from the Italian National Institute of Statistics (ISTAT) data warehouse through an annual census survey updated on 1 January 2021. According to these data, unsurprisingly, the cities with more inhabitants are likely to contain more daily positive cases. Consequently, to enable the comparison between the exposure of cities to contagion, these absolute data have been converted into relative data by dividing the total number of positive cases by the number of inhabitants in each city. Subsequently, this number has been multiplied by 1000 to reach a more understandable indicator. In other words, the number of daily positive cases for 1000 inhabitants is calculated in every 1.181 municipalities. Afterwards, these relative daily data have been transformed into annual data by simply dividing

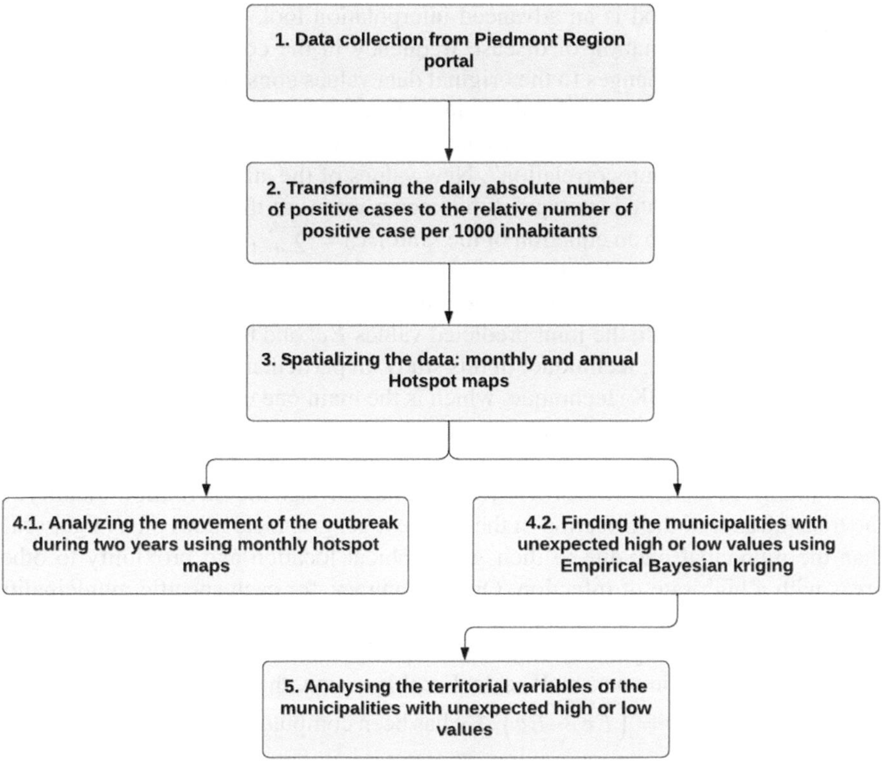

Fig. 2.2 Methodological workflow

the overmentioned indicator into the number of days each year. The output of this step was two sets of relative annual exposure to contagion (AEC_R) for every municipality during the first or second year of the pandemic. These data then became spatialized by matching the ISTAT code for each municipality to the shapefile provided by Piedmont Region Geoportal.

The first set of analysis dealt with the creation of 24 monthly and two annual maps of hot spots using the hot spot analysis tool available in Spatial Statistics Tools of ArcGIS. This tool operates by examining each characteristic in the light of its surrounding features. A feature must have a high value and be surrounded by enough features that have high values in order to be a statistically significant hot spot. The hot spot analysis helps to recognize the movement of the outbreaks within the territories.

Moreover, this study takes a closer look to the clusters of municipalities and finds out that some municipalities unexpectedly show a significantly different level of contagion compared to the surrounding municipalities. In order to find and locate these municipalities, in each of the two years, the data regarding the relative contagion have been interpolated by using the Empirical Bayesian Kriging (EBK) technique, which is available in ArcGIS Spatial Analyst Tools.

The Kriging method is an advanced interpolation tool, validated by Kyriakidis in 2004, for the estimation of disease frequency in the country scale (Kyriakidis 2004) that provides changes to the original data values considering the values in the neighbourhoods of the point and their mutual spatial correlation, i.e., by evaluating the geographic variation of the attribute in terms of what is commonly known as the "variogram" or "autocorrelation". New values of the attribute are thus obtained by computing a weighted average of the known values of the attribute in the neighbourhood according to an equation of the kind $\tilde{E}c_0 = \sum_{i=1}^{N} \lambda_i \tilde{E}c_i$, where $\tilde{E}c_i$ is the predicted value at the ith location, λ_i is the weight attributed to the predicted value at the ith location, and c_0 is the prediction location (see, e.g., Waller and Gotway 2004). How to compute the joint predicted values $\tilde{E}c_i$ and how to define the weight depends on the kriging technique; in this study, in particular, we used the Empirical Bayesian Kriging (EBK) technique, which is the main one used and, with respect to other alternative kriging techniques, enables a continuous representation of the data by predicting values in a spatial term between sample points where observations are not available. In fact, the values of AEC_R obtained through the EBK allow identifying the transmission of the infection in the different geographical areas on a larger scale than the municipal one due to their geographical location and proximity to other areas with a high rate of infection. On the contrary, for each specific municipality, the difference between the observed value of AEC_R and the predicted one through EBK highlights the singularity of the municipality with respect to its proximity, thus providing information on a small scale. For this reason, the relative difference index *RDc* defined as $RDc = \left(Ec - \tilde{E}c\right)/\tilde{E}c$ has been computed for each municipality *c*, where Ec is the observed AEC_R while $\tilde{E}c$ is predicted AEC_R through EBK. Note that the difference has been further divided by the predicted value to let comparable the differences where real AEC_R are not comparable. Thus, the relative difference index provides a measure of AEC_R with respect to the nearby cities, and for this reason, it has been used to group the municipalities in those having "high relative AEC_R" (the 25% having the higher *RDc*) and those having "low relative AEC_R "(the 25% having the lower *RDc*, which result to be negative in most of these cases). A statistical analysis of the characteristics of the two groups, in terms of indicators associated with each city, has been finally developed; graphical representations of the distributions of such indexes and suitable statistical test to compare the two groups have been performed.

The three indicators used in this statistical analysis, for which comparisons among the two groups have been performed, are the population density (D_p), the ageing index (I_a), and the annual average traffic flow ($AADT$). D_p is calculated simply by dividing the population of each municipality by the total area of each city (km2) (2021). The I_a is a dynamic statistical indicator used in demographic statistics to describe the weight of the elderly population in each population and is measured by dividing the number of inhabitants equal or more than 65 years by the population who have equal or less than 14 years. Finally, the $AADT$ represents the total volume of highway or road vehicle traffic for a year divided by 365 days. Table 2.1 presents the description of all four indicators, how they are calculated, and their data resource.

Table 2.1 Table of indicators

Indicator	Definition	Formula	Variables	Unit	Resource of data
Exposure to COVID-19 contagion (annual indicator)	The number of positive cases in each municipality relative to its population and calculated as an annual average	$AEC_R = \frac{\sum_i p_i}{I \times 365}$ $i = 1, 2, \ldots 365$	$AEC_R =$ level of exposure $P_i =$ number of positive cases in each day of the year $I =$ inhabitants of the municipality	n	The data regarding the daily positive cases are extracted from the official website of Piedmont Region. https://www.regione.piemonte.it/web/covid-19-mappa-piemonte
Population density	Measurement of population per unit area	$D_p = \frac{N}{A}$	$D_p =$ population density, $N =$ the total population $A =$ the land area covered by the population	n./square metre	Population data are extracted from The Italian National Institute of Statistics (I.Stat) on 1 January 2021. http://dati.istat.it/Index.aspx?lang=en&SubSessionId=5585f6f0-9527-4c1d-8b05-666cf0882d59# The land area covered by the population is extracted from The Italian National Institute of Statistics. https://www.istat.it/it/archivio/156224
Ageing index	A dynamic statistical indicator used in demographic statistics to describe the weight of the elderly population in each population	$I_a = \frac{P_a \geq 65}{P_a \leq 14}$	$I_a =$ ageing index $P_a =$ population age	n./n	Population data are extracted from The Italian National Institute of Statistics (I.Stat) on 1 January 2021. http://dati.istat.it/Index.aspx?lang=en&SubSessionId=5585f6f0-9527-4c1d-8b05-666cf0882d59#

(continued)

Table 2.1 (continued)

Indicator	Definition	Formula	Variables	Unit	Resource of data
Annual average daily traffic	The total volume of vehicle traffic of a highway or road for a year divided by 365 days	$AADT = \frac{1}{n}\sum_{k=1}^{n} VOL_K$	AADT = annual average daily traffic VOL_k = daily traffic on kth day of the year n = number of days in a year (365 or 366)	n./year	Rapporti sulla Mobilità Veicolare in Piemonte: https://www.regione.piemonte.it/web/temi/mobilita-trasporti/sistema-informativo-regionale-dei-trasporti/rapporti-sulla-mobilita-veicolare-piemonte

For each year, the distributions of the indicators of population density (D_p), *ageing index* (I_a), and annual average daily traffic (*AADT*) in the two groups of "high relative contagion (AEC_R)" and "low relative contagion (AEC_R)" cities have been compared to verify if there are statistically significant differences between such groups in terms of the characteristics of density, population age, and intensity of traffic of the corresponding municipalities.

2.3 Results and Discussions

As mentioned earlier, this study first analyses the contagion territorial changes and outbreak movements within the whole Piedmont Region. Subsequently, we analyse the municipalities which unexpectedly have higher or lower contagion by statistically testing three territorial indicators, including population density, ageing index, and annual average traffic flow.

The mapping of contagion over time is useful to identify and locate the hot spots of the outbreaks and to understand how the outbreak moves within the territory. In addition, it can be evaluated how the mitigation and recovery measures are affecting the territory.

As can be seen in Fig. 2.3, the first outbreak is associated with high relative contagion in the eastern territories bordering Lombardy and Emilia-Romagna Regions, where the first clusters of positive cases were found. During July and August, a significant reduction in the number of positive cases led to the absence of significant hot spots or cold spots. Unlike the first outbreak, the routes for spreading during the second outbreak originate from the southern regions of Liguria and the French Provence-Alpes-Côte d'Azur. This might be due to the mass movements of the people to the coastal regions during the summer. The third outbreak started with clusters of hot spots in the province of Torino, and until May 2021, it continued spreading and reached the southern regions. Again in the second year, during July and August

2021, a significant reduction has been noticed like the first year, and once more, the last analysed outbreak originated from the cluster of hot spots close to the southern municipalities bordering Liguria. In addition, unless the first outbreak, the eastern regions were most of the time among the cold spots, which might be due to the regional lockdowns put into place. This pattern can also be seen if the annual maps of the first and second 12 months are compared.

In addition to the analysis of changes in the contagion pattern within the territories, the scale of our analysis also allowed to take a closer look to the situation of particular municipalities as well. In more detail, although analysing the hot spots and cold spots helps to understand how the outbreaks move within the territories, there are some

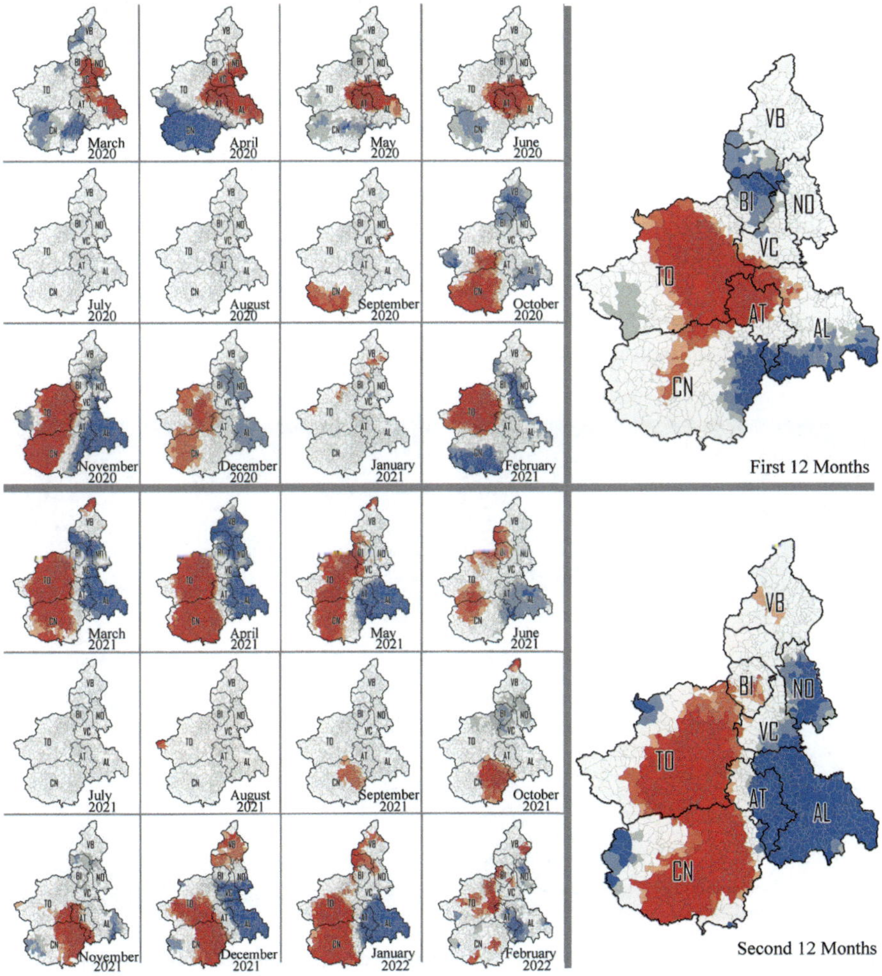

Fig. 2.3 Hot spot analysis of the COVID-19 contagion in the Piedmont Region (monthly maps are presented on the left and the annual maps are presented on the right)

municipalities that—despite being within the same cluster of hot spots or cold spots—show a significant difference in terms of contagion level regarding the neighbours. These municipalities might show either an unexpectedly higher or lower contagion. In the following phase of the study, we aim to verify if this difference is related to some territorial variables by statistically testing the relationship between the contagion and three indicators: Population density, annual average traffic flow, and ageing index.

In order to analyse the territorial characteristics of the municipalities with unexpectedly higher or lower contagion, it is first necessary to locate these municipalities. To this aim, as can be seen in Figs. 2.4 and 2.5, two maps are created and compared. The map on the left shows the level of COVID-19 contagion for each municipality using the real data. The map on the right is created using the EBK in which the value for each location on the map is affected by the nearby locations and this leads to the creation of a continuous map in comparison to the previous map which was more discrete. By using the EBK technique for each municipality, a new value is predicted based on the surrounding values, which represents the contagion considering the geographical context. Then this predicted value for each municipality is extracted using the EBK. Subsequently, this value is compared to the previous map, which represents the actual value of the contagion for 1000 inhabitants. Accordingly, the municipalities with relatively high or relatively low differences between the actual and predicted values are found. Afterwards, the higher and lower quartiles are extracted. The importance of these municipalities comes from the fact that, regarding their geographical location, their contagion rate is unexpectedly high or low, and the reason behind this fact is worth studying.

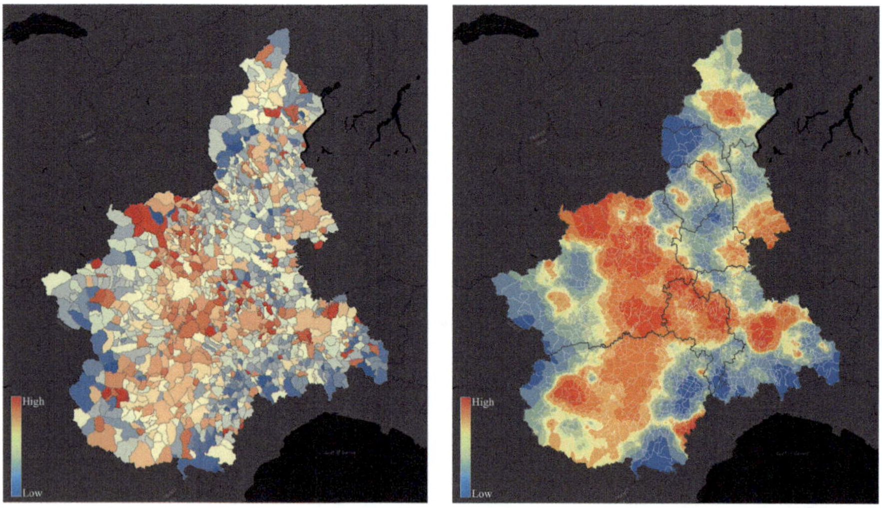

Fig. 2.4 Geographical exposure to COVID-19 contagion during the first 12 months of the pandemic using actual data (left) and EBK technique (right)—These two maps are compared to find the unexpectedly high or low value municipalities

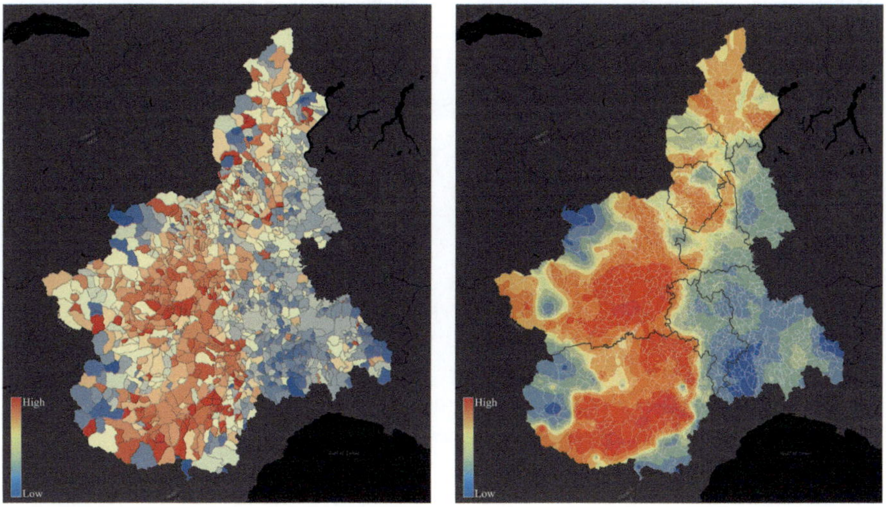

Fig. 2.5 Geographical exposure to COVID-19 contagion during the second 12 months of the pandemic using actual data (left) and EBK technique (right)—These two maps are compared to find the unexpectedly high or low value municipalities

The same procedure is also done for the data of the second 12 months of the pandemic. The actual value and EBK maps are presented in Fig. 2.5.

In addition to the AEC_R values, the territorial indicators, including D_p, I_a, and $AADT$ are also spatialised in correlation with the indicator of COVID-19 contagion, which are presented in Fig. 2.6. These maps are created using a multivariate visualization technique available in ArcGIS. The variations in the colour always represent different levels of COVID-19 contagion for the whole period of 24 months in different municipalities, while the dimension of the symbols refers to each of the three indicators. These maps are created to give an overview of the correlation between the spatial distribution of COVID-19 contagion and the spatial distribution of each indicator. It can be seen from the maps that the pattern of distribution for population density and the annual average of the daily traffic flow show a concentration of red colour big symbol size in the central zones, which suggest that high number of contagions overlaps high values for population density and traffic flow in these zones. However, this correlation is not observed for ageing index. Each of these relations will be separately tested using a set of statistical techniques to uncover their relation in the following paragraphs.

The statistical analyses are aimed at investigating the characteristics (expressed through the indicators of density, ageing, and daily traffic) of the municipalities that have very high relative AEC_R difference index (therefore with very high AEC_R compared to proximity) or very low (very negative, therefore with very low AEC_R compared to proximity). This has been done by comparing the distributions of the indicators in the two groups. The analysis has been repeated twice, once for each year of observation.

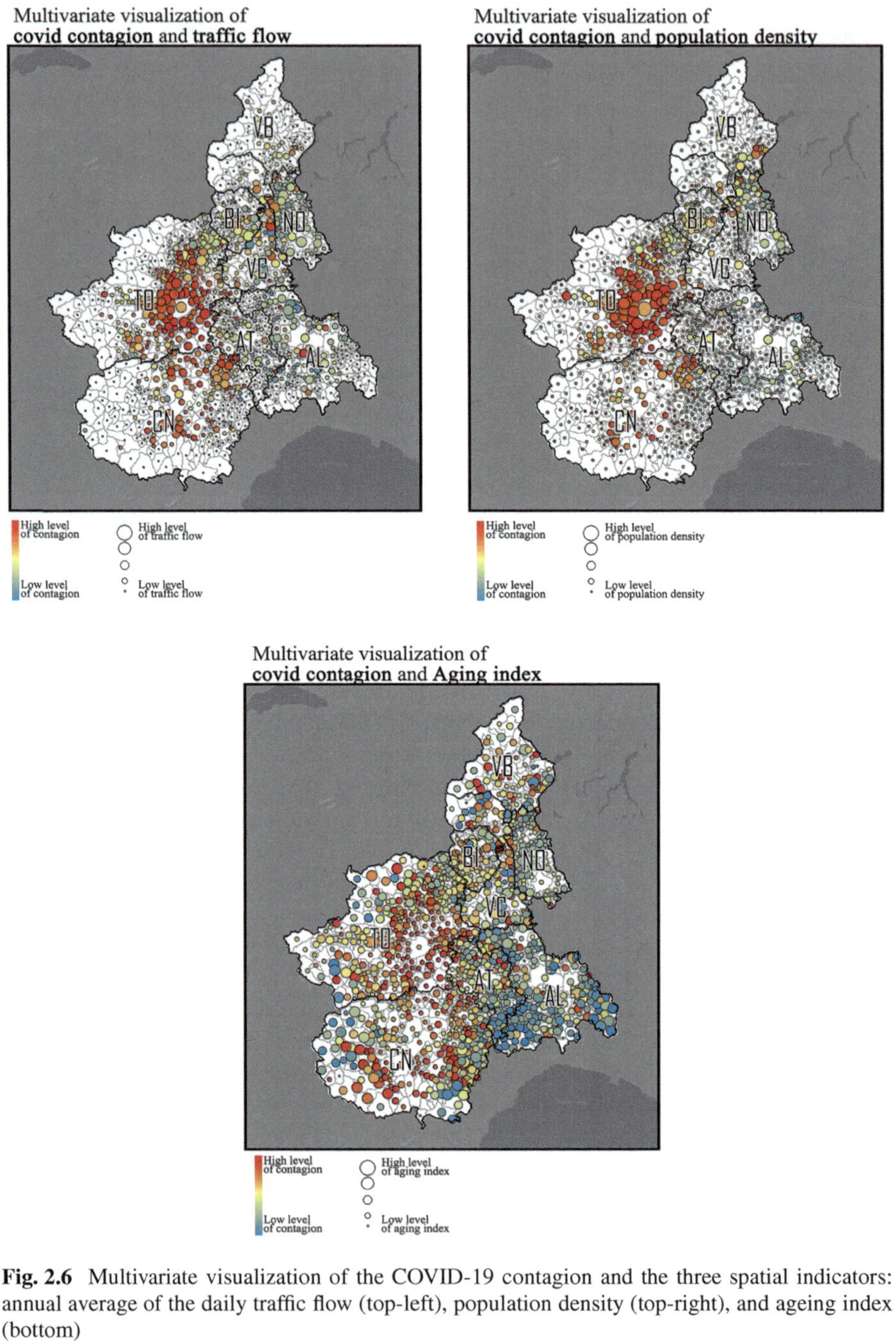

Fig. 2.6 Multivariate visualization of the COVID-19 contagion and the three spatial indicators: annual average of the daily traffic flow (top-left), population density (top-right), and ageing index (bottom)

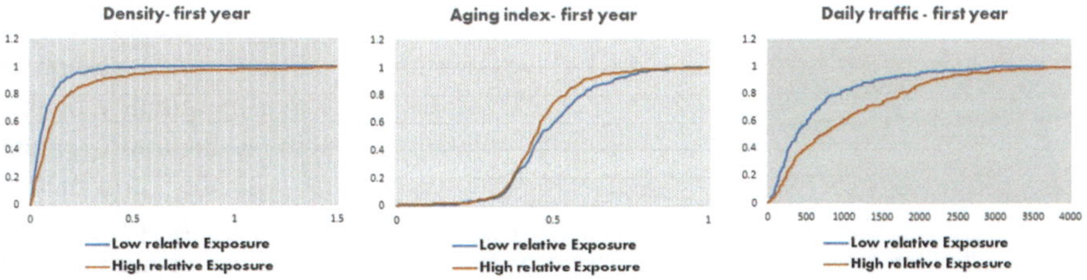

Fig. 2.7 Graphical representations of the cumulative distributions of each indicator for the municipalities with low/high relative exposure difference index during the first year of the pandemic

For the first year (March 2020–February 2021), the graphical representations of the cumulative distributions are those shown in Fig. 2.7, where the red line refers to high relative AEC_R municipalities, while the blue line refers to low relative AEC_R municipalities.

As one can see from Fig. 2.7, for all three indicators, the two distributions are stochastically ordered, i.e., they do not cross each other. This means that, for any value of the indicator, the percentage of municipalities having AEC_R smaller than such a value is always higher for the same one of the two groups. For the D_p, given any value d, the percentage of municipalities having D_p smaller than d is higher in the low relative AEC_R, (blue line) which means that this group has more "lower D_p" municipalities than the high relative AEC_R group (red line). The same happens for what concerns $AADT$, while the opposite is verified for the I_a.

A two independent samples test having equality of distributions as a null hypothesis has been performed to understand better if a testing procedure statistically confirms this graphical result. In particular, since the normality of these distributions is clearly not satisfied, the non-parametric Mann–Whitney U test has been applied (see, e.g., Fay and Proschan 2010), obtaining the p-values described in Table 2.2.

Table 2.2 Whitney U test results regarding the three territorial indicators for two groups of cities with high and low exposure during the first year of the pandemic

Mann–Whitney U test for independent samples—First year				
Indicator	Group	Mean	St. error of mean	P-value for the hypothesis of same distribution
Density index	Low relative exposure	0.071	0.004	0.00
	High relative exposure	0.163	0.017	
Ageing index	Low relative exposure	0.486	0.009	0.04
	High relative exposure	0.453	0.008	
Daily traffic index	Low relative exposure	314	21.6	0.00
	High relative exposure	501	26.8	

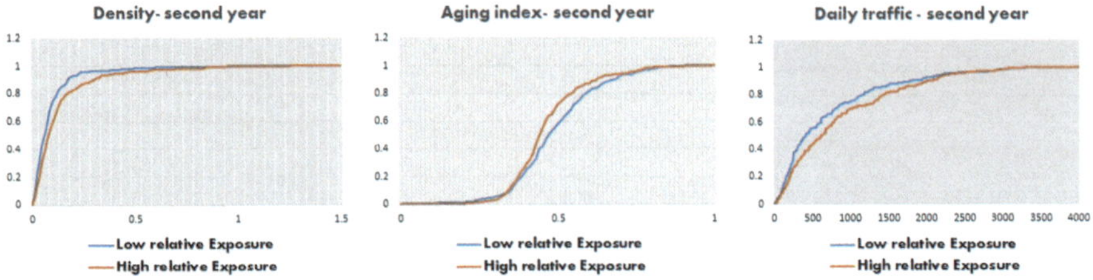

Fig. 2.8 Graphical representations of the cumulative distributions of each indicator for the municipalities with low/high relative exposure difference index during the second year of the pandemic

Equality of the distributions is strongly rejected in all three cases. One can thus affirm that municipalities in the low relative AEC_R group have, in distribution, lower D_p, low $AADT$ and higher I_a in the population.

For the second year (March 2021–February 2022), the graphical representations of the cumulative distributions are those shown in Fig. 2.8. The same results for the first year are obtained, even if the distances among curves are now smaller.

The results for the Mann–Whitney U test having equality of distributions are shown in Table 2.3. Again, the hypothesis can be strongly rejected, except for what concerns the I_a for which the *p*-value is greater than 0.1.

The results of the overmentioned analysis are in line with the hypothesis originating from mapping results, especially for the first year. It is demonstrated that the municipalities with an unexpectedly higher rate of infection than the nearby cities during the first year are more likely to have more D_p, more $AADT$ and less I_a. For the second year, two main differences are observed. The first relates to the I_a where its correlation with high exposure to the COVID-19 transmission could not be verified anymore. Secondly, for D_p, and $AADT$, the curves representing the difference between the two groups are closer to each other. These results suggest that in the

Table 2.3 Whitney U test results regarding the three territorial indicators for two groups of cities with high and low exposure during the second year of the pandemic

Mann–Whitney U test for independent samples—second year				
Indicator	Group	Mean	St. error of mean	P-value for the hypothesis of same distribution
Density index	Low relative exposure	0.081	0.006	0.00
	High relative exposure	0.122	0.008	
Ageing index	Low relative exposure	0.484	0.008	0.11
	High relative exposure	0.467	0.007	
Daily traffic index	Low relative exposure	349	21.8	0.01
	High relative exposure	448	25.6	

beginning phase of the pandemic, the impact of these two territorial variables is more significant and is more important to be considered.

2.4 Lessons Learned from COVID-19. The Leading Role of Spatial Dimension

It has been over two years since COVID-19 became a pandemic, and its behaviour is still unclear despite the health emergency ends. What is increasingly evident is that the spatial aspect of the pandemic is instead the central issue in understating the disease distribution patterns over territories. Although COVID-19 is full of unknowns, the spatial analysis assumes a leading role in interpreting the pandemic and its impact on decision-making and everyday life.

By mapping the spatial dynamics of COVID-19 transmission in the Piedmont Region through a GIS-based geostatistical analysis, this chapter initially showed that the outbreaks spread in the territories with different patterns of movements over time. These patterns are likely to be related to the implemented policies in place during the pandemic. For example, this study shows that although the first outbreak originated from hot spots of contagion in the western municipalities bordering regions of Lombardy and Emilia Romagna, during the rest of the pandemic, these municipalities turned to cold spots, probably because of the regional lockdown measures. The two annual maps to track the movement of the disease outbreaks over time performing a GIS-based hot spot analysis show how and where municipalities with either high or low values of contagion cluster spatially. In other words, to be a statistically significant hot spot, a feature must have a high value and be surrounded by additional features with high values. The hot spot analysis—performed in different periods—helped us to clarify the territorial pattern of the outbreaks during the pandemic and clarifies how the existing policies and measures affected the situation. Although proper, hot spot analysis only tells us some things about spatial contagion differences. Many times, within a cluster of municipalities which shape a hot spot (or cold spot), there are single municipalities which show a significant difference from the rest of the municipalities in the cluster. An example is shown in Fig. 2.9, which represents the data for the second 12 months of the pandemic in two different maps. The first map on the left represents a hot spot analysis. And the second one shows the heat map of the number of contagions for 1000 inhabitants in each municipality. As seen from the map on the right, some municipalities, despite being located within clusters of hot spots or cold spots, significantly differ from the nearby municipalities.

Accordingly, after performing the hot spot analysis, this study applied the Empirical Bayesian Kriging (EBK) to find and geographically locate the municipalities which show an unexpectedly high or low contagion compared to the surrounding. We found out that some municipalities—even if they shape a cluster of hot spots or cold spots with their neighbours—have unexpectedly higher or lower values of contagion regarding the nearby cities. Therefore, these municipalities were detected

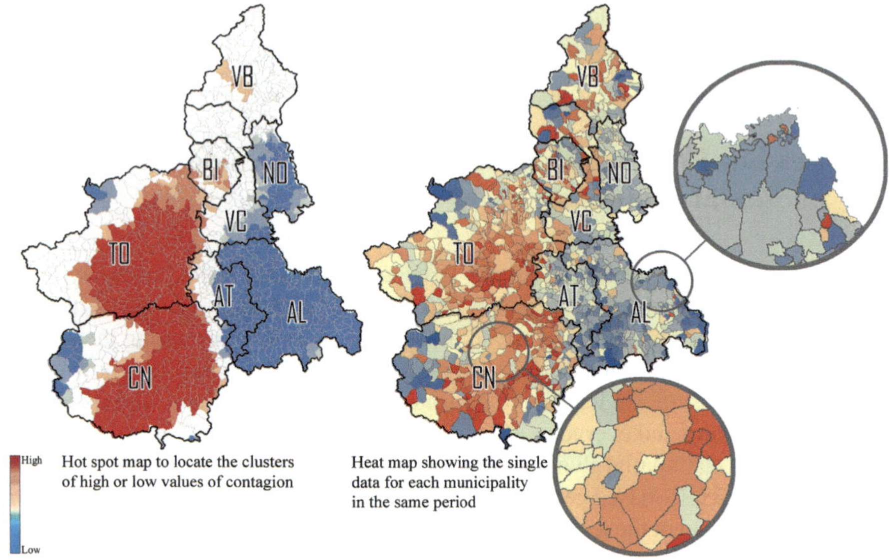

Fig. 2.9 Second year contagion map in the Piedmont Region (hot spot map on the left, and the number of positive cases for 1000 inhabitant for each municipality on the right)

using the EBK technique, and a statistical test was performed to investigate their territorial characteristics. Three indicators were explored for these municipalities. It has been found that the municipalities with suddenly higher exposure to contagion are more likely to have more population density, more mean annual traffic flow, and less ageing index, especially during the first year. The same but a slighter correlation is verified for population density and traffic flow during the second year, while the impact of the ageing index in the second 12 months of the pandemic is not verified. In a nutshell, the spread is not "nonlinear".

Moving from the analysis to the policy side, the findings of this study have several potentially significant implications for policymaking. As became clear during these two years of COVID-19, the policy responses to pandemics have a territorial dimension. Although most policy responses are at the national level and with national coverage, they result in very different regional situations once restrictive measures have been put in place. Some regions and municipalities faced more intense and/or longer-lasting consequences than others. More detailed, the current safety measures in Italy are merely planned and imposed based on the administrative borders of the regions and/or municipalities without considering the spatial variations of the phenomenon within territories. A comprehensive revision in the safety measures according to these characteristics may be helpful to cope with similar future scenarios towards more efficient actions based on real-time data and territorial analysis. For example, stricter regulation of activities and movements in areas with high levels of connectivity may be more effective at slowing the spread of COVID-19 than focusing on limiting movement everywhere. In addition, the temporal analysis shows that the hot spots and cold spots of COVID-19 contagion move over time, so future safety measures must consider the time scale to determine when and where to interfere with mitigating the impacts and enhancing territorial resilience. This issue becomes more significant considering the importance of anticipation and early action in case of such a phenomenon.

A limitation of this study regards the fact that here we analysed the spatial dimension of the pandemic considering the exposure of territories to the spread of the disease. A further study could also assess the vulnerability of the population to the phenomenon—as counterpoised to transformative resilience—so that a more comprehensive understanding is given, which directs us to analyse the territorial resilience of the studied area (Brunetta et al. 2019). From this perspective, it is pretty shared among urban scholars that the COVID-19 pandemic can become a catalyst for urban change (Zgórska et al. 2021) by supporting solutions that would not take room during different times.

Overcoming this streamlined perspective, this chapter proposes to move towards a territorial approach to pandemics aimed at tackling the challenges regarding territorial vulnerabilities, fostering the coping capacity of the territorial systems as a whole in a transformative resilience perspective. In this line, transformative resilience implies a "build back better" approach, providing the ability to cross thresholds and move territorial systems into new basins of attractions following emergent and often unknown development trajectories.

2.5 Contributions

The article has been read and approved jointly by the four authors. In particular, Ombretta Caldarice authors the final version of the paragraph 1, Franco Pellerey authors the final version of the paragraph 2, Danial Mohabat Doost authors the final version of the paragraph 3, Grazia Brunetta authors the final version of the paragraph 4.

References

Allain-Dupré D, Chatry I, Michalun V, Moisio A (2020) The territorial impact of COVID-19: managing the crisis across levels of government. OECD Policy Responses to Coronavirus (COVID-19), 10, 1620846020-909698535. https://www.oecd.org/coronavirus/policy-responses/the-territorial-impact-of-covid-19-manageing-the-crisis-across-levels-of-government-d3e314e1/

Avishai B (2020) The pandemic isn't a black swan but a portent of a more fragile global system. The New Yorker. https://www.newyorker.com/news/daily-comment/the-pandemic-isnt-a-black-swan-but-a-portent-of-a-more-fragile-global-system

Balducci A (2020) I territori fragili di fronte al Covid. Scienze Del Territorio, SPECIAL ISSUE: Abitare il territorio al tempo del Covid/Living the territories in the time of Covid: 169–176. https://doi.org/10.13128/sdt-12352
Bhadra A, Mukherjee A, Sarkar K (2021) Impact of population density on Covid-19 infected and mortality rate in India. Modeling Earth Syst Environ 7:623–629. https://doi.org/10.1007/s40808-020-00984-7
Brunetta G, Ceravolo R, Barbieri CA et al (2019) Territorial resilience: toward a proactive meaning for spatial planning. Sustainability 11:2286. https://doi.org/10.3390/su11082286
Brunetta G, Caldarice O (2020) Spatial resilience in planning: meanings, challenges, and perspectives for urban transition. In: Leal Filho W, Azul A, Brandli L, Özuyar P, Wall T (eds) Sustainable cities and communities. Springer, Cham, pp 1–12. https://doi.org/10.1007/978-3-319-95717-3_28
Buja A, Paganini Id M, Cocchio S et al (2020) Demographic and socio-economic factors, and healthcare resource indicators associated with the rapid spread of COVID-19 in Northern Italy: An ecological study. PLoS One 15(12). https://doi.org/10.1371/journal.pone.0244535
Carozzi F, Provenzano S, Roth S (2020) Urban density and COVID-19. IZA Institute of Labor Economics, Bonn, pp 1–27
Coppola A (2021) Looking through (and beyond) a truly total territorial fact: notes on the pandemics and the city. Territorio 97:7–13
Dettori M, Deiana G, Balletto G, Borruso G, Murgante B, Arghittu A, Azara A, Castiglia P (2021) Air pollutants and risk of death due to COVID-19 in Italy. Environ Res 192:110459
Fay MP, Proschan MA (2010) Wilcoxon-Mann-Whitney or T-test? on assumptions for hypothesis tests and multiple interpretations of decision rules. Statistics Surveys 4:1–39. https://doi.org/10.1214/09-SS051
Hamidi S, Sabouri S, Ewing R (2020) Does density aggravate the COVID-19 pandemic?: Early findings and lessons for planners. J Am Plann Assoc 86(4):495–509. https://doi.org/10.1080/01944363.2020.1777891
Kadi N, Khelfaoui M (2020) Population density, a factor in the spread of COVID-19 in Algeria: statistic study. Bull Natl Res Centre 44:1. https://doi.org/10.1186/s42269-020-00393-x
Kyriakidis PC (2004) A Geostatistical framework for area-to-point spatial interpolation. Geogr Anal 36:259–289. https://doi.org/10.1111/J.1538-4632.2004.TB01135.X
Mollalo A, Vahedi B (2020) GIS-based spatial modelling of COVID-19 incidence rate in the continental United States. Sci Total Environ 728(2020):138884
Piemonte R (2018). Rapporti sulla Mobilità Veicolare in Piemonte https://www.regione.piemonte.it/web/temi/mobilita-trasporti/sistema-informativo-regionale-dei-trasporti/rapporti-sulla-mobilita-veicolare-piemonte. Accessed 7 Mar 2022
Regione Piemonte (2020) Covid-19: la mappa del Piemonte https://www.regione.piemonte.it/web/covid-19-mappa-piemonte. Accessed 23 Feb 2021
Quammen D (2012) Spillover: animal infections and the next human pandemic. WW Norton & Company
The Italian National Institute of Statistics (2021a). Popolazione residente al 1° Gennaio 2021a: Piemonte. http://dati.istat.it/Index.aspx?QueryId=18540. Accessed 7 Mar 2022
The Italian National Institute of Statistics (2021b). Principali statistiche geografiche sui comuni. https://www.istat.it/it/archivio/156224. Accessed 7 Mar 2022
Waller L, Gotway C (2004) Applied spatial statistics for public health data. Wiley
Wong DWS, Li Y (2020) Spreading of COVID-19: density matters. PLoS One 15. https://doi.org/10.1371/JOURNAL.PONE.0242398
Zgórska B, Kamrowska-Załuska D, Lorens P (2021) Can the pandemic be a catalyst of spatial changes leading towards the smart city? Urban Planning 6:216–227. https://doi.org/10.17645/up.v6i4.4485

Chapter 3
The Role of the Minor Hydrographic System in Increasing the Ecological Network

Luigi La Riccia and Stefano Ferraris

Abstract This contribution describes the definition of the structure of the local ecological network. It was carried out as part of the support activities for the construction of the new urban plan of Mappano (Turin, Italy). The knowledge of the minor hydrographic system in Mappano allowed the construction of the structural map of the local ecological network, which contemplates the structural elements of the network (primary ecological network), the contiguous portions to the structural elements (areas with ecological functionality areas where it is a priority to intervene to increase ecological network), the areas of possible expansion of the network, i.e., areas with residual ecological functionality. However, there it is possible to carry out interventions useful for the protection of habitats and species of interest for conservation of biodiversity. Peripheral strips and connecting corridors, consisting of minor water canals, have therefore made it possible to better define the areas of possible expansion of the network: wetlands and marshes, in these relevant areas, represent stepping-stones of fundamental importance for rest and reproduction of many species and which need to be safeguarded in the design of the new local urban plan.

Keywords Local ecological network · Minor hydrological system · Mappano · Green and blue infrastructures · Ecosystem services

L. La Riccia (✉)
DIST—Interuniversity Department of Regional and Urban Studies and Planning, SDG11LAB, Politecnico di Torino, Turin, Italy
e-mail: luigi.lariccia@polito.it

S. Ferraris
DIST—Interuniversity Department of Regional and Urban Studies and Planning, LabFlux, Università degli Studi di Torino, Turin, Italy
e-mail: stefano.ferraris@unito.it

G. Brunetta et al. (eds.), *Post Un-Lock*, The Urban Book Series,
https://doi.org/10.1007/978-3-031-33894-6_3

3.1 Introduction: The Ecological Network for Planning at the Local Scale

Among the lessons that come to us from the health crisis from COVID-19 pandemic there is one, of fundamental importance, which concerns biodiversity. The destruction of natural habitats and the pursuit of increasingly intensive models of agriculture, breeding, and industry undermine the resources necessary for the well-being, health, and economy of human communities and at the same time expose us all to serious biological risks. The dramatic crisis of biodiversity today is therefore one of the greatest challenges that deserves attention even for health reasons (Hooper 2005). It is in this sense that the massive reaction to the pandemic crisis implemented by the European Union and largely represented by the "Next Generation EU" programme[1] should be read. This programme, confirming and adopting the sustainability of the Green Deal principles, demonstrates that the European Commission and the EU Member States, albeit through a difficult path of convergence, have at least in theory understood that it is really time to change and that the economic recovery must closely match environmental resilience and sustainability (PNRR, 2021).

Since the 1970s, when the strong pressure of economic growth and demographic expansion of cities began to show in a concrete way, there was a need for mankind to find harmony with nature (United Nations Conference on the Human Environment, Stockholm Conference 1972): climate and global changes, land consumption, deforestation, intensive agriculture, pollution, and ecological fragmentation are contributory causes of the disappearance of biodiversity (UNEP, 1992) and the destruction of ecosystems which are accompanied by the degradation of ecosystem services. The main challenges perceived by people today, in terms of risk and impact, are environmental ones (World Economic Forum 2020).

The reduction of biodiversity damages the resilience of natural systems and favours the transmission of pathogens from animals to humans (zoonoses) (IUCN 2020, 2021). Many scientific data support that the emergence and re-emergence of zoonotic diseases are linked to the unnatural coexistence between wild animals and humans, as well as to the alterations of ecosystems and the subtraction of natural habitats from wild species due to uncontrolled urbanization (IPBES 2018; IPCC 2019).

The COVID-19 pandemic has highlighted the vulnerability to the reactions of nature and the poor ability to mitigate their impacts (preparedness), with serious damage to health, social cohesion, and socio-economic well-being. It is therefore all the more important to account for current and potential economic costs, through a correct quantification of ecosystem services, deriving from the degradation of natural assets: the positive externalities for biodiversity and ecosystems must certainly be valued. Urban green space networks, together with natural and semi-natural ecosystems around cities, allow urban areas to be more sustainable and tackle many challenges including air pollution, noise, heat waves, hydrogeological instability and

[1] https://europa.eu/next-generation-eu/index_en.

a better management of the water cycle, conservation of the resource through the strategy of green and blue infrastructures.

Therefore, since the seventies, urban planning practices demonstrated the potential of ecological network to contribute to challenges such as health, species protection, biodiversity protection, and climate change adaptation (Benedict and McMahon 2002). When understood as part of local ecological network, these and other emerging challenges and trends must be considered not just as obstacles to overcome but as important drivers for investing the future urban planning choices (La Riccia 2015, 2017).

The identification of the most suitable planning scale to trigger, starting from an ecosystemic vision, territorial policies aimed at the design of ecological networks is a question strictly connected to the definition of the concept of "local", which cannot univocally coincide, according to a common name widely used in urban planning, with only the municipal planning area (Selman 2006). The ecological network in fact refers to an open system of territorial relations between the different biological and landscape elements that constitute it and cannot, therefore, be enclosed and delimited within strictly defined administrative limits. Thus, involving variously localized portions of the territory, the ecological network interacts with multiple scales and territorial planning tools. The urban planning scale that comes closest to the methodological perspective outlined for an adequate planning and management of ecological networks therefore seems to coincide with that represented by municipal and park territorial planning, which today have a more direct operation and a higher capacity of integration.

Ecological networks can be framed among planning strategies including an articulated set of territorial actions aimed at mitigating the effects of environmental fragmentation of anthropogenic origin at all levels of ecological organization. The main objective of this type of planning is, therefore, the conservation of biological diversity and dynamic processes that allow the maintenance of vitality and functionality over long periods of biological populations and communities, ecosystems, and landscapes.

Being a set of territorial actions that refer to environmental sustainability policies, acknowledging EU and international guidelines, the priority given to nature conservation actions, beyond the intrinsic value attributed to biodiversity and the need for its protection, it implies a series of positive consequences also on a human level. These consequences can be of a social, cultural, aesthetic-perceptive nature, being the interventions, in general, aimed at improving the environmental quality and conserving resources, and their usability, for future generations.

To deal with the design of an ecological network, it is important to have a cognitive framework relating to the basic ecological and landscape disciplines inherent to this problem (Antrop 2001, 2004): the models of population structure and dynamics, the ecology of biotic communities, the ecology of the landscape, the study cultural and visual perceptive landscape, land uses, and conservation biology.

3.2 The Minor Hydrological System in Mappano

The current form of settlement of Mappano (Turin, Italy) reflects the nature of a territory whose use has been fundamentally agricultural for centuries, dotted with some large historic farmhouses. Over the centuries, in fact, exclusively agricultural activity has developed, with the creation of a complex system of irrigation and reclamation canals (to make land that is too humid and characterized by stagnant water usable) which has evolved over time, with a texture of minutely branched and mutually interconnected streams. On this territorial matrix, the recent urbanization and infrastructure of the twentieth century has superimposed itself in a disorderly form, creating significant hydraulic problems. In most of the territory, the aquifer is located a few metres from the ground level and is exposed to high vulnerability.

The area is not crossed by main natural waterways, so there are no relevant elements of the fluvial dynamics characteristic of large waterways. As part of the drafting of the new local urban plan of Mappano, a very accurate campaign of land surveys was carried out aimed at representing the system of canals and ditches in the various functional and critical aspects. The articulated system of irrigation canals, the vast agricultural area innervated by them, the large development of rows of trees still present, and the stretches of water, probable residues of mining activities of the past, constitute a real network of significant size (see Fig. 3.1), so much so as to be an important piece of the ecological network on a territorial scale, as well as a qualification element from a landscape point of view.

Some problems from a hydraulic point of view are evident: the ecosystem and hydrogeological issue connected to the Stura di Lanzo Torrent is actually relevant for the flows that cross Mappano. The surface water system has promoted a historical productive function in the Mappano area, giving rise between the nineteenth century and the first half of the twentieth century to laundry activity. It still finds its present in some active industrial laundries.

Today, the drinking water catchment wells, located mainly to the north of the town, are strategic, especially if related to the issue of hydrogeological risk (see, for example, the intense atmospheric events that occurred in 2019) and require special attention to areas at risk of flooding. If we focus our attention on the system of surface waters coming from the Stura river, we can point out in particular the Bealera Nuova (with its spillway), the San Giorgio Canal, and the foothill spillway channel protecting Leinì-Settimo (the so-called Scolmatore Ovest), which are already the subject of coordinated interventions in relation to the hydrogeological safety of the territory.

Fig. 3.1 Minor hydrographic system of canals in Mappano area (*Source* La Riccia, SDG11Lab)

3.3 The Value of Minor Waterways for the Construction of Local Ecological Network and Green and Blue Infrastructures

The waterways, even minor ones, have a very high value for the local ecological network: the water flow constitutes a natural line of continuity; the banks of the waterways and the lateral bands also present intrinsic impediments (topographical and related to flood events) for the construction of buildings and works of various kinds; for these reasons, it is along the waterways that, even in heavily man-made areas, residual elements of naturalness are more easily found. Moreover, these are particular elements of naturalness, endowed with specific ecosystemic characteristics (hygrophilous and aquatic facies, sheltered environments with high slopes, very often not representative of the surrounding areas), necessary but not sufficient to express the multiple needs of an ecological network. It is a complex category within which it is possible to distinguish further cases:

Main ecological river corridors or similar to be strengthened and/or rebuilt for multiple purposes. It is the set of main waterways that can form the backbone for multi-purpose (ecological and fruitful) redevelopment projects of a certain breadth;

Minor watercourses of significant ecological value (ichthyofauna, aquatic life in general, naturalistic requalification of the bank vegetation) or belonging to complex or development-relevant minor water systems, for which a priority policy of maintenance and enhancement can be proposed of biological resources.

If we consider the ecological reticularity of the territory, we understand that, in the case of Mappano, it rests on this very dense network of minor waterways and therefore represents the backbone of this fragile system of residual habitats, precisely because this territory is predominantly agricultural.

The construction of an ecological reticularity derives from a functional approach that not only guarantees the protection of ecosystems but also their efficiency and, therefore, the possibility of fulfilling their ecosystem functions and related ecosystem services, a vital aspect of the anthropic use of natural resources. The objectives of the elaborations carried out are: (1) to analyse the state of naturalness and biological diversity at different scales (from wide area to local scale); (2) to prioritize the pursuit of ecological coherence in spatial action; (3) to protect areas relevant to the conservation of ecosystems from the effects of potential impacts from external human activities; (4) to restore degraded ecosystems; (5) to promote the sustainable use of natural resources, compatible with the protection of biodiversity and naturalness.

The map in Fig. 3.2 shows the network of minor waterways superimposed on the structure of the local ecological network: it is the result of the combination of two methodologies (Voghera and La Riccia, 2016): ARPA (Piedmont Region, 2015) and ENEA (LGRE, Metropolitan City of Turin, 2014–2016). The ARPA methodology (DGR no. 52-1979 of 31/07/2015), which identifies the elements of the ecological network on the basis of faunistic and vegetation indicators and modelling tools in order to identify the areas of ecological value and ecologically permeable areas of the territory. The reference framework of indicators is based on the Piedmont

Land Cover (level IV) and the respective classification according to the EUNIS system (EEA, 2007). This classification allows a score to be applied to the following systemic classes (mammals, avifauna, invertebrates), which is based on a species-habitat affinity analysis. To this is added an analysis of the vegetation aspects on the basis of four indicators (distance from the climax, naturalness, degree of floristic biodiversity, conservation importance), resulting in areas of ecological value for vegetation. Ecological Value Areas (AVEs) are then identified. The ENEA methodology identifies the ecological function (Voghera et al. 2020) of the territory, starting from the different types of land use on a Piedmont Land Cover basis (level IV) and the criteria for assessment, defining five key indicators for evaluating the ecological status: naturalness, conservation relevance, extroversion, fragility, and irreversibility. The integration of the results of the two methodologies results in the map of the structurability of the ecological network. This map shows the systems constituting the local ecological network (REL) consisting of three main elements: (1) structural elements of the network (primary ecological network), i.e., areas of high and moderate ecological functionality as well as areas hosting point conservation emergencies, i.e., of significant naturalness and relevance for biodiversity conservation; (2) portions contiguous to the structural elements (buffer 50 m), i.e., the areas with residual ecological functionality in which it is a priority to intervene to increase the functionality of the primary ecological network and for which to implement protection measures to maintain the primary ecological network; (3) areas of possible expansion of the network, i.e., the areas with residual ecological functionality, but on which it is possible to implement interventions useful for the protection of habitats and species of interest for the conservation of biodiversity.

3.4 Discussion

The analyses carried out on the Mappano territory highlight several vulnerabilities from different points of view: firstly, the intensive use of resources must be considered, in particular land consumption due to settlement sprawl phenomena, with the modification of the dynamics of minor canals and water use; secondly, from the hydrological point of view, a scarcity of hydrological functions can be noted, with the presence of artificial water sources and distributions, the reduction of surface, ground and underground water supply.

Consequently, the sensitivity of the water ecosystem to climate change. Hydraulic works on the river network have led to morphological change, river silt dynamics and management, and lowering of the riverbed.

According to CICES classification (v. 5.1),[2] it is possible to select some specific ecosystem services considered important from the point of view of the minor hydrographic system in relation with local ecological network (Table 3.1).

[2] https://cices.eu/content/uploads/sites/8/2018/03/Finalised-V5.1_18032018.xlsx.

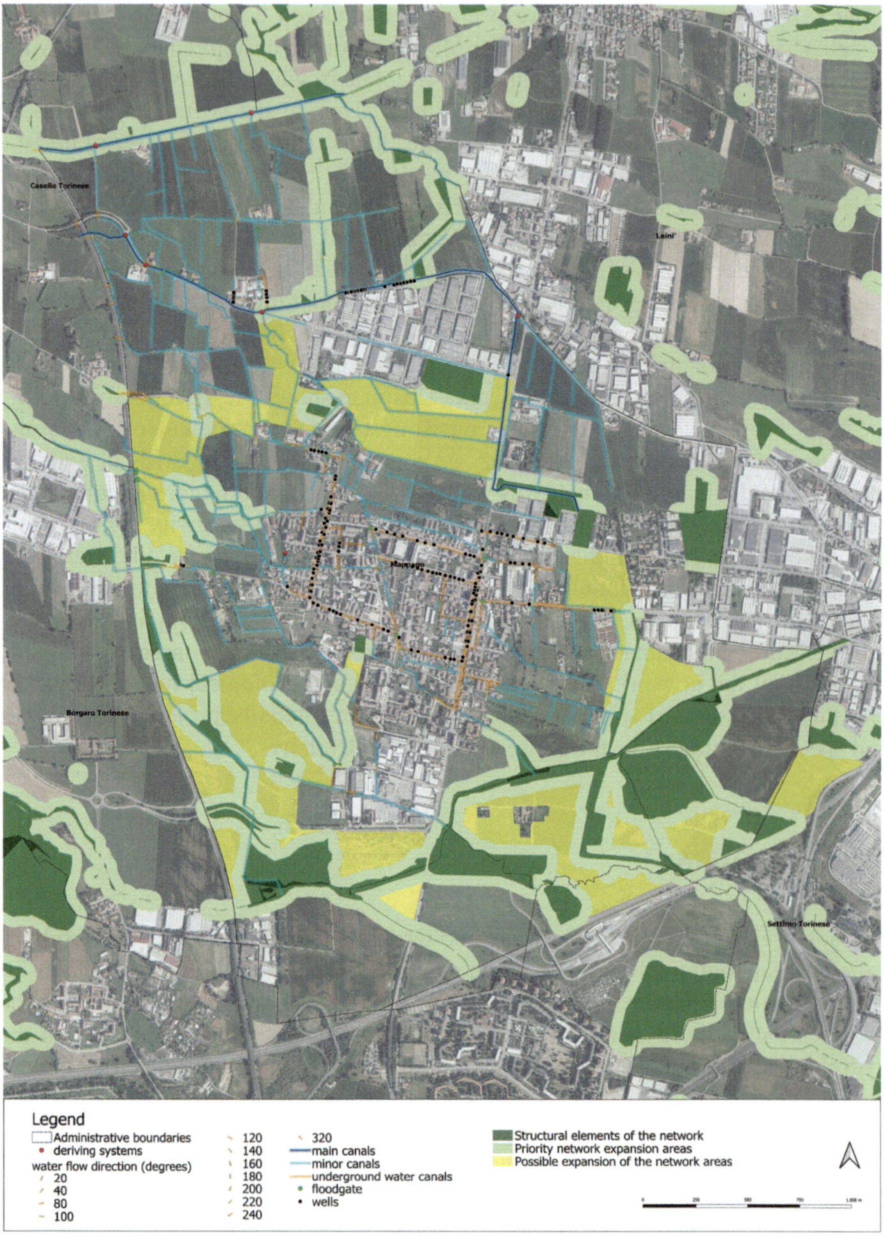

Fig. 3.2 Relation between the minor hydrographic system of canals and the structure of local ecological network in Mappano (*Source* La Riccia, SDG11Lab)

Table 3.1 Selection of ecosystem services (ES) important from the point of view of minor hydrographic system and landscape in Mappano

	Supporting	Regulating	Provisioning	Cultural
ES responding to vulnerabilities	Maintaining population and habitat (including gene pool protection)	Surface water filtration by ecosystems Regulating of the chemical condition of freshwater by living processes Hydrological cycle and water flow regulation (including flood control and shores protection)	Surface water for drinking with minor or no treatments Ground and subsurface water for drinking Surface water use for non-drinking purposes Grassland biomass: cultivated terrestrial plants (including fungi, algae) for nutritional purposes	Spirituality and religion, cultural diversity, inspiring creativity and art Cultural heritage, sense of belonging, social relations Environmental education Mental and physical health, well-being, leisure, recreation and ecotourism, aesthetic service Existence and legacy value Noise attenuation, visual screening Increased urban landscape value
Other ES supporting resilience		Bio-remediation by micro-organisms, algae, plants, and animals Filtration / sequestration / storage/ accumulation by micro-organisms, algae, plants, and animals	Grassland biomass: fibres and other materials from cultivated plants, fungi, algae and bacteria for direct use or processing (excluding genetic materials) Reared Animals link to agricultural sector preservation (existing farming animals) Fibres and other materials from reared animals for direct use or processing (excluding genetic materials)	

Ecosystem services, adequately assessed, must be preserved and increased in order to support green and blue infrastructures, to respond to vulnerabilities and support territorial resilience. In these terms, it is important to maintain the naturalness of water ecosystems, including riparian areas and vegetation, wetlands and meadows, and connecting them with the agri-ecosystem elements.

The idea is to strengthen the water network as a structural element of Mappano landscape, improving and guarantee river functionality. The re-naturalization of minor waterways could increase the ecosystem services responding to vulnerabilities, specifically supporting and regulation ones (increasing water ecosystem multifunctionality and diversity), also using environmental engineering techniques.

3.5 Conclusion

An ecological network project that aims to interact effectively with the other networks that make up the territory (settlement and infrastructural) must adopt a series of elements that are part of a complex territorial reality. The main biodiversity reservoirs are given by the areas in which the natural environment has characteristics of high extension, of differentiation of the habitats, of continuity between the ecosystem units. Areas of this type (comparable to large, basically continuous "core areas") are still present but are basically disappearing due to the strong anthropogenic presence.

Minor waterways, including ponds and small lakes, low-level streams, streams, ditches and springs, are the most numerous freshwater environments globally, are fundamental to the biodiversity of freshwater, and are increasingly recognized for their role in sustaining ecosystem services (Giudice et al. 2023). Despite this, minor waters remain the least studied part of the water environment and are largely excluded from water management planning. We identify research priorities to support better protection of small waters and recommend policy actions needed to better integrate small waters into watershed and landscape management. Key requirements are identifying reliable monitoring programmes for minor waterways, developing effective measures to protect biodiversity and the ecosystem services they provide, and ensuring that regulators take full account of this critical part of the water environment.

The margins of natural matrices can be sharp or, as often happens, frayed. In the event that contiguous portions to the structural elements are in direct contact with the more anthropized territories but still incorporates significant presences of natural and water units, these can play significant basic roles of support for possible recolonization of the anthropized territory by species of interest.

With a view to reconstructing a functional ecological network, it is necessary to distinguish the units capable of constituting, in terms of size and articulation, an ecosystemic cornerstone capable of self-sustaining, from the connecting elements whose role is above all to favour biotic movements on the territory. Within highly anthropized areas, these cornerstones take on the configuration of real functional nodes, whose spatial definition depends on the connection objectives and the current natural presences.

In order to be able to speak of an "ecological node", it is necessary that a sufficient quantity of spatially close natural elements overall exceed a certain dimensional threshold, so that a "critical mass" is built up, capable of providing sufficient habitats for maintaining stable populations of species of interest, as well as allowing a differentiation of internal habitats capable of improving conditions for biodiversity purposes. To complete the primary nodes, other areas can be identified (areas of possible network expansion) to which an ecological function with the role of steppingstone can be attributed. Strengthening the natural capital present on the territory, even outside the main network, also the establishment of an intermediate steppingstone could support point where the connections would be too long.

References

Antrop M (2001) The language of landscape ecologists and planners—a comparative content analysis of concepts used in landscape ecology. Landsc Urban Plan 55:163–173

Antrop M (2004) Landscape change and the urbanization process in Europe. Landsc Urban Plan 67:9–76

Benedict MA, McMahon ET (2002) Green infrastructure: smart conservation for the 21st century. Sprawl Watch Clearinghouse Monograph Series, Washington DC

EEA (2007) EUNIS Habitat Classification, EEA, European Environment Agency, EUNIS terrestrial habitat classification review 2021_1

Giudice B, Novarina G, Voghera A (2023) Framing green and blue infrastructure. In: Giudice B, Novarina G, Voghera A (eds) Green infrastructure. The urban book series. Springer, Cham, pp 1–11. https://doi.org/10.1007/978-3-031-28772-5_1

Hooper DU et al (2005) Effects of biodiversity on ecosystem functioning: a consensus of current knowledge. Ecol Monography 75:3–35

Intergovernmental Science-Policy Platform on Biodiversity and Ecosystem Services [IPBES] (2018) Land degradation, human-wildlife contact, and zoonotic spillover. Assessment report on land degradation and restoration, Chap. 5.4.2. Accessed 17 Feb 2022. https://ipbes.net/sites/default/files/2018_ldr_full_report_book_v4_pages.pdf

Intergovernmental Panel on Climate Change [IPCC] (2019) Risks to humans from disrupted ecosystems and species. Special report on climate change and land, Chap. 7.3.3. https://www.ipcc.ch/site/assets/uploads/sites/4/2019/11/10_Chapter-7.pdf. Accessed 17 Feb 2022

International Union for Conservation of Nature [IUCN] (2020) Statement on Covid 19 Pandemic. https://www.iucn.org/news/secretariat/202004/iucn-statement-covid-19-pandemic. Accessed 17 Feb 2022

International Union for Conservation of Nature [IUCN] (2021) Resilience is in our nature. IUCN World Conservation Congress, Marseille, 3–11 Sept

La Riccia L (2017) Landscape planning at the local level. Springer, Cham

La Riccia L (2015) Nature conservation in the urban landscape planning. In: Gambino R, Peano A (eds) Nature policies and landscape policies. Towards an alliance. Springer, Dordrecht, pp 157–164

Piano Nazionale di Ripresa e Resilienza [PNRR] (2021) https://www.governo.it/sites/governo.it/files/PNRR.pdf. Accessed 17 Feb 2022

Piedmont Region (2015) Legge regionale del 29 giugno 2009, n. 19 Testo unico sulla tutela delle aree naturali e della biodiversità. Approvazione della metodologia tecnico-scientifica di riferimento per l'individuazione degli elementi della rete ecologica regionale e la sua implementazione. http://www.regione.piemonte.it/governo/bollettino/abbonati/2015/36/attach/dgr_01979_680_31072015.pdf. Accessed 16 Feb 2022

Selman P (2006) Planning at the Landscape Scale. Routledge, New York

United Nations Environment Programme [UNEP] (1992) Convention on Biological Diversity. Rio De Janeiro

Voghera A, Giudice B (2020) Defining a social-ecological performance to prioritize compensatory actions for environmental regeneration. The experimentation of the environmental compensation plan. Sustainable cities and society, pp 1–15, ISSN: 2210-6707. https://doi.org/10.1016/j.scs.2020.102357

Voghera A, La Riccia L (2016) Landscape and ecological networks: towards a new vision of sustainable urban and regional development. Laborest 12:89–93

World Economic Forum (2020) The global risk report 2020. https://www.weforum.org/reports/the global-risks-report-2020. Accessed 17 Feb 2022

Chapter 4
Ecosystem Services and Territorial Resilience: The Role of Green and Blue Infrastructure

Carolina Giaimo, Benedetta Giudice, Giulio Gabriele Pantaloni, and Angioletta Voghera

Abstract Responding to the new environmental, ecological, and social emergencies requires a shift in strategies and urban design models. In the contexts of sustainability and resilience, green and blue infrastructure (GBI) is a wide-ranging concept that can help overcome the usual dichotomies of urban growth versus green or the built environment versus nature. This provides different benefits, both environmental and ecological and social and economic. In urban contexts, green spaces play a strategic role due to the number of typologies and functions that vary from neighborhood spaces to green, play, and sports facilities to protected areas of territorial scale. In this way, the planning and design of GBI take on the triple objective of regenerating fragile and degraded ecosystems from an environmental, social, and economic point of view. Focusing on this assumption, we describe how the GBI that develops along the axe of the Stura di Lanzo river in a multiscalar mosaic of soils at both local and territorial levels can determine options for the ecosystem quality of the metropolitan area of northern Turin. We suppose that mapping ecosystem services (based on a correct land use/land cover design) can support designing new urban and regional plans to improve resilience.

Keywords Ecosystem services · Green and blue infrastructures · Spatial planning

C. Giaimo (✉) · B. Giudice · G. G. Pantaloni · A. Voghera
DIST—Interuniversity Department of Regional and Urban Studies and Planning,
R3C—Interdepartmental Responsible Risk Resilience Centre, Politecnico di Torino, Torino, Italy
e-mail: carolina.giaimo@polito.it

G. Brunetta et al. (eds.), *Post Un-Lock*, The Urban Book Series,
https://doi.org/10.1007/978-3-031-33894-6_4

4.1 Introduction

4.1.1 The Territorial Interpretation for Resilience and Well-Being

Since vulnerability to pandemics, environmental and economic crises, social disaggregation, and climate change-related impacts are increasing, above all in EU urban areas that are home to over two-thirds of the population,[1] biodiversity, green and blue infrastructure (GBI), and ecosystem services (ES) are gaining importance for resilience and sustainability. Moreover, in line with sustainable development goals (UN 2015), these concepts are crucial to avoid landscape and environmental trivialization, degradation of nature and social inequality, and enhance well-being.

In this changing context, we focus on ES, GBI, and biodiversity through the adoption of a specific analysis methodology (see Sect. 1.2) within a particular case study, the Stura di Lanzo river and the "Basse di Stura," an urban park in the northern part of the city of Turin. This case study is an excellent example of where different landscape features coexist and where different scenarios can be imagined. Indeed, this case helps us understand how to contrast ecosystem degradation in cities, providing a wide range of ES (TEEB 2010) and developing the vital functions for social direct or indirect benefits in relation to the post COVID-19 situation.

Our application is reinforced by plenty of global and European strategies and policies (such as the EU Biodiversity Strategy 2030 and the Post-2020 Global Biodiversity framework) and scientific debates that stress the attention on the necessary transition toward sustainability and resilience. Indeed, these policies support enhancing biodiversity (giving a central role to protected areas—Pas—and other green areas), ecological reticularity and ecosystem functionality, regulating climate, ensuring health and well-being, purifying water and air, maintaining soil fertility, and ensuring species reproduction. The interconnection of PAs and green areas allows for building a multiscalar and multifunctional GBI, intended as an open system of relationships between these different green spaces. For its characteristics, this system is the most appropriate approach toward sustainability and resilience (Voghera and Giudice 2019) in planning and design, integrating different levels, scales, and types of policies and plans from the territorial context to local and sectorial plans and projects. Moreover, due to their multifunctionality, green spaces within GBI play an important social role, bringing people closer to nature. Additionally, in this system, ES, PAs, and biodiversity are the building blocks of GBI, where rivers and green spaces represent the corridors for guaranteeing connectivity, encouraging the provision of ES and associated benefits to humans, and integrating biodiversity in planning and ecological design at different scales.

Piedmont, the region of our case study, has a multifaceted system of PAs and green spaces (national parks, natural parks, provincial protected areas, natural reserves,

[1] Considering global data, today, more than 4.3 billion people live in urban areas: this means over half of the world (55% in 2017) live in urban settings (Ritchie and Roser 2018).

natural safeguard areas, and special reserves, sites of Natura 2000 network, Unesco WHS, and Biosphere Reserves). Furthermore, the topic of interconnection between different green areas is one of the central objectives of the policies and strategies promoted by the Metropolitan City of Turin (CMTO), in line with the indications defined at the national scale by the National Strategy for Biodiversity (MATTM 2010) and in the perspective of supporting policies to control land take. In particular, the CMTO developed in 2014, in collaboration with ENEA and the Politecnico di Torino (DIST), the Guidelines for the Green System (LGSV), which includes the Guidelines for the Ecological Network (LGRE), which identify the Provincial Ecological Network and provide municipalities with general regulatory guidance to control land take, increase, qualify, and conserve ES, with a focus on biodiversity and the promotion of rational use of natural resources.[2] Furthermore, GBI, ecological networks, and PAs have become central in planning tools promoted by the CMTO: the Metropolitan Strategic Plan (in particular, the Axis no. 2 "Green Revolution and Ecological Transition") and the preliminary technical proposal of the Metropolitan General Territorial Plan on GBI, ES, and PAs that incorporates and implements the Regional Ecological Network.

4.1.2 Methods and Data

Geospatial data can play a key role in supporting more resilient urban planning. Accurate and timely geospatial data, along with the tools needed to convert them into meaningful information for decision-making, can be strategic for better knowing and planning GBI with greater awareness of the value of ES.

Interoperable, high-quality, and timely geospatial information and analysis are fundamental prerequisites for good policymaking. This is particularly evident when there is a need to integrate both quantitative and qualitative information from different sources and, often, different methodologies. Instead, the lack of sufficient, reliable, high-quality, and timely geospatial information leads to inconsistent and incorrect decisions or even non-decision-making.

In particular, if we focus on ES, there are now many projects and initiatives that, both at territorial and local scales, reason with them and their ability to support land-government decision-making processes, aimed above all at the good use and proper management of the soil resource (Nedkov et al. 2018; Burkhard et al. 2018) through the design of green frames. In Italy, the most relevant recent experiments are often accompanied by urban planning processes (EU Life projects SAM4CP 2014–2018, SOS4Life 2016–19) and territorial planning (see Province of Turin 2014).

In the PostUnlock project, we chose to deepen the assessment of the habitat quality (HQ) ES, considered one of the most significant and structural ES to describe ecosystem functionality (Assennato et al. 2018) and, therefore, also the resilience of

[2] Many interesting local experimentations applied this methodology in some pilot municipalities of Piedmont (Bruino, Ivrea-Bollengo, Chieri, Mappano). See Voghera et al. (2017).

territories. Recognizing the level of quality of habitats at the urban scale is relevant because the interactions between living organisms and the physical environment give rise to functional relationships that characterize different ecosystems, ensuring their resilience, their maintenance in a good state of conservation, and the provision of ES (ISPRA-CATAP 2012). For this reason, the issue is becoming increasingly relevant within large European cities, as the ability to adapt to climate change is strongly linked to the state of ecosystems and the biological diversity they contain; the greater the degree of biological diversity, the greater the ability of species to adapt to the new living conditions produced by climate change (MATTM 2010) and to positively affect the well-being of urban communities. In addition, the 2016 OECD report highlighted the correlation between cardiovascular and respiratory diseases with the increased presence of fine particulate matter in urban settings, with a consequent increase in the economic cost of health expenditures, putting the focus on those supporting and regulating ES that are characterized by indirect human demand (improving air quality, CO2 absorption, etc.) but of broader collective interest and to the base of human life.

The proposed reflection is transcalar and transdisciplinary. In the experimentation conducted on the territorial context of Basse di Stura, the ecosystem analysis is the outcome of the application of two methodologies: (i) the InVEST software[3] assessment model that uses geospatial data that best describes the relationship between mapping ES and essential human life needs (Costanza et al. 1997; MEA 2005), (ii) the ENEA's bioecological model (Provincia di Torino 2014) that offers an in-depth and fertile reading of land uses to assess the bioecological value and structure of ecosystems. Since both models have some limitations, depending on the diversity of approaches and computational algorithms as well as the availability and usability of input data (including implications due to the 2021 update of the Piedmont Land Cover data), the paper presents a broader reflection regarding the effectiveness and reliability of the assessment conducted.

4.2 Morphological Conditions and Land Uses in Basse di Stura

As already mentioned, the case study we selected is the area of Basse di Stura,[4] located on the northern edge of Turin, in District 5 (Fig. 4.1). The local land use plan (*Piano Regolatore Generale*—PRG) of 1995 identifies it as Urban and River Park (P.17).

---

[3] Software developed within "The Natural Capital" project by Stanford University, University of Minnesota, The Nature Conservancy, and the World Wildlife Fund.

[4] The site has been the subject of different studies in the context of joint activities between the DIST Department of Excellence (Call 2017) "*Ecowelfare e governance intercomunale. Il suolo come infrastruttura per la rigenerazione dei territori*" (under the direction of prof. C. Giaimo) and Valium (directed by M. Bottero). See Giaimo (2020a, b, c), Giaimo et al. (2021a, b).

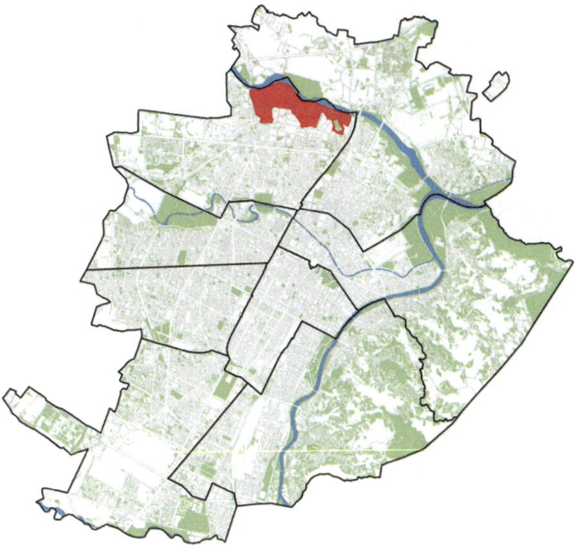

Fig. 4.1 Localization of Basse di Stura within the municipal territory. Elaborated by Pantaloni G. G

With an area of about 150 ha extending for almost 3 km along the right bank of the Stura di Lanzo river, Basse di Stura lies at the edge of the urbanized area, just south of Turin's northern freeway. Here, the Stura river once flowed in a typically agricultural landscape, traces of which can still be found thanks to the presence of some farmsteads (notably "La Ressia," "Il Canonico," or "Boscaglia," and "La Carpegna") and some agricultural land interlocked in a context strongly affected by the harmful effects of the functions and activities settled within the area and in its proximity. Indeed, Basse di Stura is surrounded by a set of viable (including the freeway) and technological infrastructure networks such as the AMIAT landfill (historically among the largest in Italy), also designated as a fluvial urban park on the opposite bank of the river. Currently, thanks to a series of remediation works, the Marmorina Park has been created in place of the old landfill.

Within Basse di Stura, included for many years in the Ministry of the Environment's list of Italian most polluted industrial sites (which has financed part of its safety), a set of impactful activities were located, such as heavy industries (Teskid), incinerators (Stureco), quarries now abandoned, industrial dumps (a former solfatara), and gravel extraction activities. Basse di Stura was successively downgraded as a site of regional interest by a Decree of the Ministry of the Environment. Moreover, its implementation is delayed by the need to provide for essential remediation works[5] as well as the permanence of some still active activities located on the site.

[5] The PRG subordinates any interventions to the preparation of an Environmental Recovery Executive Plan for the entire area—to be submitted for evaluation and authorization by the competent bodies—that takes into account, first of all, the following conditions: (i) the works must be located in areas which are not exposed to the risk of flooding, (ii) termination of polluting activities, and (iii) reclamation of polluted areas.

Its implementation (not yet activated by current planning activities) makes Basse di Stura a fundamental "piece" to be connected to the larger system of urban and river parks (already implemented and planned). It assumes roles and values that intercept a wide sphere of functionality, disciplinary contributions, and multiple spatial scales from the local to the vast area. This strategic relevance is also evident since the area is partially included in the system of PAs of the Po river belt—Turin section—"Basse di Stura Stralcio Area." Indeed, the site, which is also part of the "Torino Città d'Acque" project, is located within an environmental and landscape system connecting high-value green areas such as La Mandria Park, Mesino Reserve (Po-Stura Confluence), the hill of Turin and Superga Park, the Lanca di Santa Marta and Stupinigi Park (Fig. 4.2).

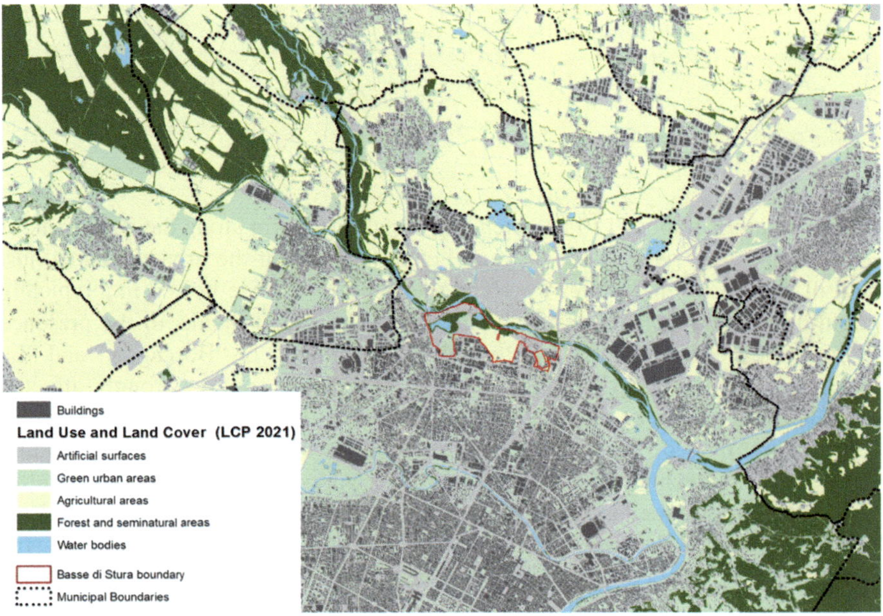

Fig. 4.2 Land use and land cover map (LCP 2021) with the location of Basse di Stura in relation to the ecological-environmental connectivities of Stura di Lanzo river. Elaborated by Pantaloni G. G

4.3 Mapping Ecosystem Service for Territorial Resilience

4.3.1 The Land Cover Piemonte Database and the Habitat Quality Model

The Land Cover Piemonte (LCP) database of 2021 is the most up-to-date and detailed open-access Land Use Land Cover (LULC[6]) database referring to the regional territory of Piedmont. The research group used it to support (i) an initial reading and interpretation of the morphological and settlement structure of the area and (ii) as the input data necessary for the proper functioning of the assessment model based both on ENEA indicators and to calculate the HQ service through InVEST (and in the SimulSoil application[7]).

In this research, one of the main advantages related to this increased geometric precision is the possibility of recognizing those urban green porosities that, although less extensive, play a fundamental role in assessing urban ecosystems.

Concerning the state-of-the-art scenario of the soils within Basse di Stura, a mosaic of uses and covers characterized by a strong unevenness emerges (Fig. 4.3): residential, manufacturing, and commercial human activities (16%), coexist with a large portion of land where there are sand and gravel mining activities (23%), arable agricultural soils, meadows, and pastures (26%) linked to the presence of the historic Martini and Ressia farmsteads, urban green areas (13%) of which it is sometimes not easy to distinguish the artificiality or the presence in the subsoil of an impermeable capping positioned to secure the aquifers. In addition to these anthropogenic activities, natural soils are composed of both spontaneous and riparian vegetation (17%) extended along the northern boundary of the area, where the Stura river flows. Finally, the two quarry lakes (5%), although artificial and polluted, constitute bodies of water with spontaneous vegetation undergoing renaturalization along the banks.

Using a LULC basis within dynamic ecosystem analysis models simulates alternative land use scenarios and allows us to observe how changing urbanized soils corresponds to a consequent change in the ecosystem performance delivered by the soil itself. In addition, such models help understand how the same quantitative soil design assumptions can generate greater or lesser impacts on ecosystem performance in the case of different physical-spatial correlations between different soils.

InVEST's HQ model combines information on land uses and land cover (derived from LULC map bases) with elements recognized as threats to biodiversity, generating habitat quality maps as outputs. Five degrees of naturalness[8] were associated

[6] LULC provides a classification of the terrestrial land that identifies (i) the type of land cover and (ii) the type of anthropogenic land use, which can be used to trace the relationships between changes in land uses and land cover and the ecosystems' capacity to deliver goods and services.

[7] Realised by the EU Life SAM4CP Project, since 2016 it has been widely used in local planning activities of the Turin and regional area.

[8] As part of the LIFE + SAM4CP Project, the values of naturalness at the national level were derived through an expert-based approach. At the local scale, the reference was the Guidelines for the Ecological Network (LGRE) which assigned a score from 1 to 5 (where higher values correspond

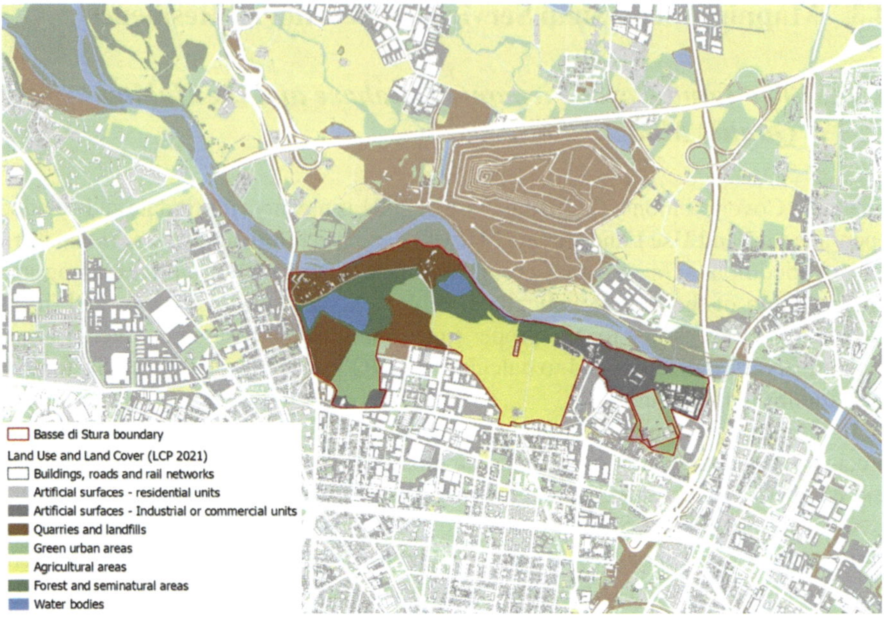

Fig. 4.3 Spatialization of LULC in Basse di Stura, Land Cover Piemonte 2021. Elaborated by Pantaloni G. G

Table 4.1 Naturalness levels derived from Provincia di Torino (2014)

1st level	*Land use typologies coinciding with climate and paraclimax stages*
2nd level	*Land use typologies coinciding with preclimatic stages*
3rd level	*Seminatural land use typologies, even if with relevant anthropogenic determinism*
4th level	*Seminatural land use typologies, even if with relevant anthropogenic determinism but not artificial*
5th level	*Land use typologies coinciding with artificial areas*

with each subcategory of land uses and land covers based on the presence/absence of anthropogenic disturbance and proximity to the climax[9] (Table 4.1) and then systematized with the spatial distribution of elements that may compromise ecosystem naturalness.

to more natural habitats), integrated with permeability values in the anthropogenic land use and land cover classes.

[9] Equilibrium situation of an ecosystem (Provincia di Torino, 2014).

The model, which takes into account 12 habitat types and considers as elements of residual or no naturalness the urbanized system, agricultural areas, and infrastructural network, allows the integration and revision of specific parameters by users, although it is necessary to keep the structure of the calculation algorithm unchanged.

Relating features placed in close proximity to each other and at a certain distance from each other introduces the concept of model dynamism, inasmuch the variables that contribute to defining the habitat quality level do not refer exclusively to the intrinsic characteristics of the single "pixel" of the soil itself (i.e., the one indicated as part of the ecosystem whose habitat quality level is to be measured), but also to parameters affecting soils placed nearby as possible sources of threat.

To assess the degree of impact of each threat, the model uses parameters such as (i) the distance (MAX DIST) between the threat source and the habitat (i.e., the maximum distance of influence that the threat exerts on habitat quality, measured in km), (ii) the decay, in space, of the threat (DECAY) of linear or exponential type, (iii) the weight (WEIGHT) of the threat and finally, referring to land uses within habitats, (iv) the sensitivity to threats.

The definition of the latter factor is done through a previously mentioned ecological sensitivity matrix, which reports the interactions between classes of land uses (to which an initial naturalness value is assigned) and threats.

The outcome of the algorithm, which relates all the variables listed above, consists of a map in which each soil pixel (with 5×5 m resolution) is assigned a habitat quality value, which, simplifying, is defined through a weighing of an initial naturalness value, related to the specific external threats that the model recognizes as detractors.

4.3.2 Design Scenarios for Basse di Stura

The application of the ecosystem assessment model described above allows for the identification of four design scenarios on the area considered ("*Parco dei Parchi,*" "*Trees,*" "*Res non Aedificatoria,*" and "*Coesistenze di Stura*") and highlights the sensitivity of ecosystem performance to changes in the mosaic of land uses (Giaimo et al. 2021b). The spatialization of biophysical values of HQ stresses that the soils with a more pronounced suitability to play the role of natural habitat are those belonging to the Stura di Lanzo river. Indeed, the river, running transversely through the urban fabric of the City of Turin, plays a role in environmental connectivity between the hilly part of the city, the La Mandria Nature Park, and beyond (Fig. 4.4).

Compared to the state-of-the-art described above, the four identified scenarios propose urban regeneration interventions that, although with different specific aims and objectives, intend to achieve a greater degree of permeability of the urban fabric and recover part of the natural riparian vegetation along the Stura di Lanzo river (Table 4.2), through soil desealing and reclamation actions. Commenting on the arrangements envisaged by the four proposals regarding the increase of permeable land covers, of particular interest are the cases of *Coesistenze di Stura* and *Res Non Aedificatoria*. While the first envisions a large presence of soils intended for an urban

Fig. 4.4 Spatialization of HQ biophysical values in the northern Turin edge with Basse di Stura and along the Stura di Lanzo river (HQ model, InVEST). Elaborated by Pantaloni G. G

park, the latter project assumes the total replacement of agricultural soils in favor of a large urban park and forest-like vegetation.

Consequently, the trend of reduction of urbanized land uses and land cover (not counting, in this specific case, soils that fall into the urban greenery category) varies depending on the design proposals, in a range from − 13% (as in the case of the Trees project) to a maximum of − 35% in Stura Coexistence, affecting the biophysical performance levels of soils differently. Concerning the design intentions briefly described, Fig. 4.5 allows for the interpretation of these urban planning choices, to make explicit a more or less pronounced correspondence between new urban development and the achievement of specific goals of improving ecological-environmental conditions and welfare.

Without undertaking a critical and analytical reading of the four scenarios, the proposals make explicit, albeit with different emphases, a strong focus on a dimension that is not only local but also large scale. Indeed, they attribute to Basse di Stura a dual role: (i) an urban standard that is today not usable by the community and (ii) a fundamental piece of the GBI system that, by territorial extension and geographic location, would allow reconfiguring the ecological-environmental profile of the urbanized northern edge of Turin and inserting itself into an environmental system of metropolitan importance.

To prefigurate future scenarios, it is fundamental to know ecosystem values to investigate the quality and critical characteristics of soils in a framework of a vast

Table 4.2 Outline of the distribution and percentage of land uses and land cover. Comparison between the state-of-the-art and design scenarios

Basse di Stura: land use and land cover	State-of-the-art (%)	Design scenarios			
		Parco dei Parchi (%)	Trees (%)	Res non Aedificatoria (%)	Coesistenze di Stura (%)
Urbanized soils—residential	1	2	5	1	1
Urbanized soils—commercial and productive	14	9	12	17	0
Roads	1	10	1	2	3
Quarries and landfills	23	0	9	5	0
Urban greenery	13	27	20	37	48
Agricultural land	26	23	22	0	20
Natural and seminatural	16	23	27	33	24
Artificial reservoirs	5	5	5	5	4
Total	100	100	100	100	100

Fig. 4.5 Standardized level of average biophysical performance of HQ. Comparison of the state-of-the-art and design scenarios on the three ES. Elaborated by Pantaloni G. G

area and to contextualize its ecological-environmental conditions. In this perspective, some recent works (Giaimo et al. 2021a) have considered Basse di Stura within a broader territory identified with the term "Northern Quadrant," with an extension equal to about 27 km^2 of the territory North-West of Turin. The analyses conducted within this area have underlined how, although the regenerative transformation of Basse di Stura is still unfinished, it has higher HQ values than the surrounding territory. This result is attributable to the greater territorial extension of the "Quadrant" and the higher presence of anthropized soils that characterize the very dense urban fabric bordering the park. These initial considerations, although carried out by taking into consideration a spatial area that does not fully meet the requirements of a Functional Ecological Unit (Santolini and Morri 2017), highlight the potential that the regeneration of Basse di Stura expresses under multiple perspectives: urban settlement quality, fruition for the community, well-being, and health.

4.4 Discussion and Open Issues

The experimentation described above well highlights how the recovery of a compromised portion of land, interclosed but located in a peripheral context of the city and in close connection with ecological and environmental reticularities of supra-local relevance, represents a concrete opportunity to pursue an improvement of the ecological-environmental conditions of the context, as well as the redistribution of spaces for the community in a marginal area and partly lacking in public services. Indeed, the rehabilitation of this territorial area assumes a double relevance, both urbanistic and ecological-environmental. On the one hand, the PRG recognizes it as an urban standard intended for urban and river parks, while on the other hand, it constitutes a fundamental component of a system of GBI (Giudice et al. 2023). More specifically, concerning urban standards, it is worth mentioning that even though the quantity provided by the PRG is largely satisfactory, the territory surrounding Basse di Stura is still partly lacking them. All these considerations make Basse di Stura a place with a high potential to provide good livability and foster new forms of interaction between people and nature.

The outcomes of the experimentations show the need to adopt an integrated and multidisciplinary approach, including analyses of urban vulnerability and social and ecological-environmental aspects that consider the future park as a piece of a larger urban framework. For example, the implementation of the Basse di Stura Park needs to be pursued within a process that involves the entire Stura di Lanzo river.

Finally, what has been argued on spatial databases and ecosystem analysis modeling underlines how crucial attention must be paid to the type of data processed within computational software (as well as the functioning of computational algorithms). The experience highlights how the descriptive content of databases, as well as their accuracy, can strongly influence the outcomes that can be obtained through the use of ecosystem assessment models, as well as the need to put in place new mechanisms aimed at constructing the necessary information, such as the biophysical parameters that support ecosystem assessment, which is no longer consistent with existing databases. Finally, it is essential to recall that ecological-environmental assessment of soils cannot be identified as the bearer of absolute and all-encompassing information, but a multidisciplinary approach to issues affecting urban regeneration processes is needed.

4.5 Conclusion

These approaches highlight the importance of evaluation methods to design a GBI-based resilience. Evaluation methods allow for measuring the ecological quality of territories, identifying territorial and local stakes, and delineating strategic, transversal, and multiscalar design actions. It is challenging to decide which method provides better support to the objectives of resilience as it depends on different factors.

Furthermore, these approaches can be used to frame and guide design solutions, redesign the quality of urban spaces at local and vast scales, and rethink post-unlock cities with new performative "standards." The new park will be a node of a multi-scalar system for ES and biodiversity valorization, expanding the resilience of the surrounding areas by transforming a quarry into a new stepstone of the ecological system of the Turin metropolitan cities.

In this perspective, measuring ecological quality and the resilience of a local system is a fundamental requirement for the selection of territories to be transformed to create an interconnected and reticular green system guaranteeing multiple equilibria and the stability of a social-ecological system by increasing and maintaining ES.

Acknowledgements The text is the result of work coordinated and shared among the authors in which paragraphs 3.1.1 is attributed to Angioletta Voghera and Benedetta Giudice; paragraphs 3.2 and 3.3 to Carolina Giaimo and Giulio G. Pantaloni; paragraphs 3.1.2, 3.4, and 3.5 bring together methodologies and reflection shared by all authors.

In addition, we would like to thank Gabriella Negrini (Dist/CED-PPN) and Luigi La Riccia (Dist/SDG11 Lab) for their support in collecting cognitive data and their contribution during the discussion of the results.

References

Assennato F, Braca G, Calzolari C et al (2018) Mappatura e valutazione dell'impatto del consumo di suolo sui servizi ecosistemici: proposte metodologiche per il Rapporto sul consumo di suolo. Rapporti ISPRA

Burkhard B, Santos-Martin F, Nedkov S, Maes J (2018) An operational framework for integrated mapping and assessment of ecosystems and their services (MAES). One Ecosystem 3:e22831. https://doi.org/10.3897/oneeco.3.e22831

Costanza R, d'Arge R, de Groot R et al (1997) The value of the world's ecosystem services and natural capital. Nature 387:253–260

Giaimo C, Lazzarini L, Pantaloni GG (2021a) Paradigmi ecosistemici per il nuovo welfare urbano. Il caso dell'area Basse di Stura a Torino. In: Ricci L et al-(a cura di) "Città pubblica e nuovo welfare. Una rete di reti per la rigenerazione urbana", Urbanistica Dossier online, INU Edizioni, Roma, p 118–125

Giaimo C, Pantaloni GG, Vitulano V, Barbieri CA (2021b) Nuove tecniche e paradigmi per la pianificazione e progettazione urbanistica. Scenari ecosistemici per l'area Basse di Stura a Torino. In: Murgante B, Pede E, Tiepolo M (eds) Innovazione tecnologica per la riorganizzazione spaziale.

Atti della XXIII Conferenza Nazionale SIU DOWNSCALING, RIGHTSIZING. Contrazione demografica e riorganizzazione spaziale, Torino, 17–18 June 2021b, vol 09, Planum Publisher e Società Italiana degli Urbanisti, Roma-Milano, p 58–65. https://doi.org/10.53143/PLM.C.921
Giaimo C (2020a) Esiti di un workshop didattico multidisciplinare. In: Giaimo C (ed) Tra spazio pubblico e rigenerazione urbana. Il verde come infrastruttura per la città contemporanea. Urbanistica Dossier online, 17, INU Edizioni, Roma, p 126–128
Giaimo C (2020b) La strategia urbanistica del verde nel PRG di Torino 1995. In: Giaimo C (ed) Tra spazio pubblico e rigenerazione urbana. Il verde come infrastruttura per la città contemporanea. Urbanistica Dossier online, 17, INU Edizioni, Roma, p. 81–86.
Giaimo C (2020c) Rigenerare lo spazio pubblico. Standard, suolo e servizi ecosistemici. In: Giaimo C (ed) Tra spazio pubblico e rigenerazione urbana. Il verde come infrastruttura per la città contemporanea. Urbanistica Dossier online, 17, INU Edizioni, Roma, p 5–8
Giudice B, Novarina G, Voghera A (2023) Framing green and blue infrastructure. In: Giudice, B, Novarina G., Voghera A (eds) Green infrastructure. The urban book series. Springer, Cham. https://doi.org/10.1007/978-3-031-28772-5_1
ISPRA-CATAP (2012) Glossario dinamico per l'Ambiente ed il Paesaggio—Revisione marzo 2012. Available via https://www.isprambiente.gov.it/files/pubblicazioni/manuali-lineeguida/mlg-78.1-2012-glossario-dinamico.pdf. Accessed 26 Jan 2023
Ministero dell'Ambiente e della Tutela del Territorio e del Mare—MATTM (2010) Strategia nazionale per la biodiversità in Italia. Available via https://www.minambiente.it/sites/default/files/archivio/allegati/biodiversita/estratto_strategia_eng.pdf. Accessed 8 November 2020
Millennium Ecosystem Assessment—MEA (2005) Ecosystems and human well-being: biodiversity synthesis. World Resources Institute, Washington, D.C.
Nedkov S, Zhiyanski M, Borisova B, Bratanova-Doncheva S (2018) Mapping and assessment of ecosystem condition and ecosystem services across different scales and domains in Europe. One Ecosystem 3:e29288. https://doi.org/10.3897/oneeco.3.e29288
Provincia di Torino (2014) PTC2 Progetto definitivo. Linee guida per la rete ecologica, Torino. Available via http://www.cittametropolitana.torino.it/cms/risorse/territorio/dwd/linee-guida-sverde/lgsv_lgre.pdf. Accessed 26 Jan 2023
Ritchie H, Roser M (2018) Urbanization. Published online at OurWorldInData.org. Retrieved from: https://ourworldindata.org/urbanization. Accessed 16 Jan 2023
Santolini R, Morri E (2017) Criteri ecologici per l'introduzione di sistemi di valutazione e remunerazione dei Servizi Ecosistemici nella progettazione e pianificazione. In: Arcidiacono A, Di Simine D, Oliva F et al (eds) La dimensione europea del consumo di suolo e le politiche nazionali. Rapporto 2017, INU Edizioni, Roma, p 149–154
TEEB (2010) The economics of ecosystems and biodiversity: mainstreaming the economics of nature: a synthesis of the approach, conclusions and recommendations of TEEB
United Nations—UN (2015) Transforming our world: The 2030 agenda for sustainable development, A/RES/70/1. Available via http://www.sustainabledevelopment.un.org. Accessed 15 Jan 2023
Voghera A, Giudice B (2019) Evaluating and planning green infrastructure: a strategic perspective for sustainability and resilience. Sustainability 11:2726. https://doi.org/10.3390/su11102726
Voghera A, Negrini G, La Riccia L, Guarini S (2017) Reti ecologiche nella pianificazione locale: Esperienze nella Regione Piemonte. Reticula 14:1–9

Chapter 5
Indicators and Scenarios for Sustainable Development at the Local Level

Alice Borsari, Patrizia Lombardi, and Sara Torabi Moghadam

Abstract Cities around the world have faced the impact of the COVID-19 pandemic with unprecedented speed, due to our hyper-connected society. As history teaches us, epidemics plague society because of the vulnerabilities generated by humans through their relationships with the environment, with other species and with each other. The recent pandemic is a stark reminder that urbanization has changed the way people and communities live, work, and interact, and it is even more necessary than in the past to adopt a multidisciplinary approach to the development of systemic operational skills that can address complex issues within cities. This work showed how many measures adopted during the emergency have now become part of daily life. The lesson of the pandemic is that people's health is connected to and dependent on the health of the planet, and cities are at the center of this relationship. The objective of the research starts from the need to identify a selection of post-COVID indicators providing an analysis methodology suitable for the creation of its own final set with the identification of specific key performance indicators (KPIs) of the project, replicable in other urban contexts, on which to base the analysis of the level of local sustainability, especially at the neighborhood scale. The proposed methodological framework is developed in two phases: (1) indicator selection and (2) baseline scenario, set out to investigate the existing correlations between the urban environment and the neighborhood level of cities. On the basis of the assessment of the KPIs, selected on the basis of numerous comparisons with the project's internal and external stakeholders, thanks to the creation of an interactive dashboard with Tableau software, it was possible to analyze the basic scenario of proximity at the neighborhood scale for the City of Turin, highlighting weak points and priority areas

A. Borsari (✉)
EURAC Research, Urban and Regional Energy Systems (URS) Team, Bolzano, Italy
e-mail: alice.borsari@eurac.edu

P. Lombardi · S. T. Moghadam
Interuniversity Department of Regional and Urban Studies and Planning, Politecnico di Torino, Viale Mattioli 39, 10125 Turin, Italy
e-mail: patrizia.lombardi@polito.it

S. T. Moghadam
e-mail: sara.torabi@polito.it

G. Brunetta et al. (eds.), *Post Un-Lock*, The Urban Book Series,
https://doi.org/10.1007/978-3-031-33894-6_5

on which to act, experimenting with the theme "Inhabiting proximity" as an urban response to the pandemic.

Keywords COVID-19 · Transformative resilience · Territorial vulnerability · Local level · Urban Health · Agenda 2030 · SDG 11 · Post-COVID indicators · Proximity · Spatial decision support system

5.1 Introduction

During the last two years, cities around the world dealt with the impact and spread of the COVID-19 pandemic with unprecedented speed, due to our hyper-connected society (Aditi et al. 2020). There is growing evidence that anthropogenic impacts on ecological systems and the different challenges which the world faces: population growth; globalization; pandemic diffusions; climate change, and the consequent environmental degradation are now the main driver of planetary change needing more adaptability and innovation to promote health and well-being in cities.

The lesson learned from the COVID-19 pandemic is that people's health is connected and closely dependent on the planet's health, and cities play an important role in this relationship (Capolongo et al. 2020). With the urban population set to increase over time, leading to a world population of 9.3 billion in 2050 (Chelleri 2012) the difficult undertaking of global sustainability rests largely on urbanization processes. Cities are defined as intricate systems consisting of social, ecological, and economic dimensions responsible for many of the trends considered unsustainable that push our planet beyond its ecological boundaries of development.

The present chapter aims to show how urbanization and the different challenges have modified the way people and communities live, work, and interact during the time and it is, therefore, even more, necessary than in the past, to adopt a multidisciplinary approach to the development of systemic operational skills that can address complex issues within cities through the use of Key Performance Indicators (KPI), and in this way, the problem exposed by the coronavirus pandemic is a perfect scenario for thinking cities in a more resilient and sustainable way.

Our understanding of cities has certainly changed, and this situation has inevitably turned out to be an unprecedented alarm bell that has highlighted the system's poor understanding of the risk's nature and interdependencies between sectors, showing more attention in the implementation of NBSs within urban contexts in response to current problems.

Since its first applications, the NBSs approach has always wanted to be seen as a way of supporting human life and activities, addressing different societal challenges in terms of multiple benefits or ecosystem services. Given the complexity of cities, the need for a healthier environment is increasingly recognized as well as the importance of connecting urban space with natural areas. The sanitary crisis has offered an opportunity to re-think the relationship between cities and nature and the possibility of channeling the urban technological transition to counteract climate change, support biodiversity, reduce pollution, and improve the well-being of inhabitants.

Today we need to develop transdisciplinary methodologies using new tools for the development of integrated measures that adequately address different economic, social, environmental, and political aspects to promote health and well-being in cities. On these front scenarios and indicators stand as means to simplify and reduce the complexity of the real world to a potentially high but limited number of factors, analyzing and monitoring their interaction aimed at supporting the formulation of policies.

5.2 Methodology

The goal of this chapter is to propose a clear and simple methodology analysis that can be replicated in all urban contexts, through the creation of a specific set of indicators, to analyze the level of local sustainability, especially at the neighborhood scale. The use of indicators within the decision-making processes allows a better understanding of the data (Angelakoglou et al. 2019) and the possibility of evaluating the expected impacts to direct future policies for the redesign of post-pandemic cities (Moghadam and Lombardi 2018).

The proposed methodological framework is developed in two different phases: *(Phase 1) Indicator selection* and *(Phase 2) Baseline scenario*, with the aim of investigating the neighborhood scale, placing the city as a new generator of public health (Fig. 5.1).

The proposed two phases allow the development of an appropriate vocabulary and lexicon aimed at analyzing the responsiveness of the territory, and in this case, the experiments were carried out for the Municipality of Turin, experimenting with the

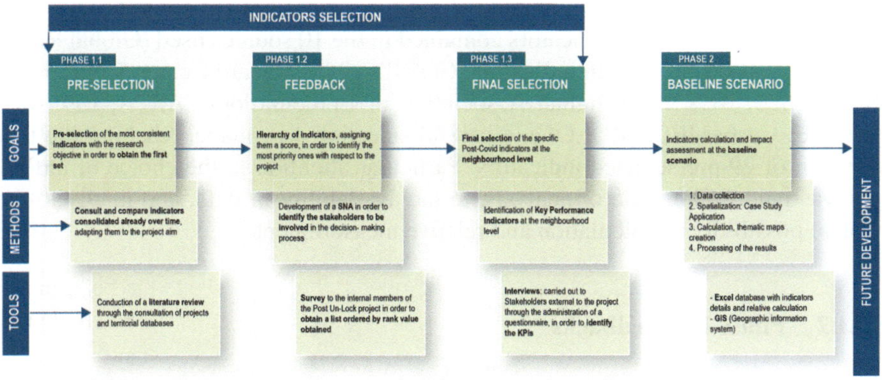

Fig. 5.1 Methodological framework

city of proximity as an urban planning pandemic response. Measuring, evaluating, and spatially visualizing the impact are the keys to the proposed evaluation process.

The proposed methodological framework intends to select and evaluate, from the set of final selected indicators, those that are defined as the Key Performance Indicators (KPI) (Genta et al. 2019), considered mandatory. To be as site-specific as possible, we started with the application of the experimentation to the neighborhood scale, especially for the City of Turin within the "Crocetta" neighborhood, the seat of the Polytechnic of Turin's University, to become a pilot project to be adapted to the whole municipal area.

The different phases of the work are presented below:

5.2.1 Phase 1: Indicators Selection

The first phase of the work has the ultimate aim of measuring the level of sustainability and local proximity on the scale of the neighborhood by quantifying the phenomena that constitute it, thus producing the data necessary for the subsequent phase, the operational phase of quantifying the impacts. The proposed phase is structured in three different steps: *Pre-selection, Feedback,* and *Final selection.*

Steps will be detailed as follows:

5.2.1.1 Phase 1.1 Pre-selection

The objective of this phase is to pre-select the indicators that are more consistent with the final objective of the research to obtain the first set. The method was to read the indicators already consolidated over time, proposed within various projects and territorial databases, adapting them to the specifications of the Post Un-lock Project, modifying their descriptions and units of measurement. Starting from the consultation of the more than 700 indicators contained in the 10 sources used (Global Reference List of 100 Core Health Indicators; COVID-19 Dashboard Center; UN Habitat; ISTAT 2020; BES 2020; Italian Association of Epidemiology; ARPA; LEGAMBIENTE; CesbaMED; MOLOC), a careful selection was made, reaching the first selection of 29 pre-selected indicators. Particular attention to the choice of indicators was paid to the scale of analysis, to the availability of data, and finally to the relative possibility of calculation and relative measurement.

5.2.1.2 Phase 1.2 Feedbacks

From the earliest stages of this methodology, it was essential to identify the Stakeholders to be involved in the decision-making process to obtain expert opinions on the subject. Within this phase, the 29 pre-selected indicators in the first phase were submitted to the expert opinion of the Stakeholders within the present Post Un-lock project, to understand the indicators that, compared to the project, were more suitable. Their involvement in the decision-making process represented a very important step as it helped to know the existing data available, to determine the relevant objectives and to propose a final common strategic vision concerning the project (Genta et al. 2019).

The objective of this phase, through the compilation of a questionnaire in Google Modules, was that of the hierarchization of the post-COVID indicators by assigning them a score from 0 to 4 ((0 = not important at all, 1 = not important, 2 = on average important, 3 = important 4 = very important) to obtain a final list ordered by the value of importance in descending order. The feedbacks obtained were a total of 13, and the indicators as a result of the further changes made in the description and the unit of measurement went from 29 to 26 since 3 were identical. The interesting aspect that emerged from the hierarchization of the values obtained for each indicator is the perfect congruence with the topic under consideration. All 26 indicators turned out to be perfectly calculable indicators at the municipal level for the City of Turin, and in particular, among these 26, 15 indicators were specific to the neighborhood scale.

Since the ultimate goal of this work is to analyze the correlations existing between the urban environment and the proximity level of cities, especially at the neighborhood scale, the last phase of the selection of indicators was to ask oneself, among the 15 indicators on the scale of the neighborhood which were the Key Performance Indicators (KPI).

5.2.1.3 Phase 1.3 Final Selection

The tool used for the validation of the final set was the conduct of interviews with Stakeholders external to the Post Un-lock Project, asking them to evaluate the importance of the 15 indicators proposed at the neighborhood scale, defining them as "accepted", "To be modified", or "rejected", giving reasons for their choices.

To conduct the interviews, some were carried out via video call, in particular to the local Stakeholders in the Turin context, which allowed the joint discussion of the results obtained and the votes assigned, and other votes were assigned through the compilation of an additional questionnaire, created on Google Modules to be distributed in a widespread manner. For this phase, the Stakeholders who provided their assessment were a total of 50, thus making the analysis as complete as possible and achieving the initial set of objectives.

The results obtained showed that all 15 indicators proposed on the neighborhood scale have been defined as "accepted", thus resulting in the specific KPIs of the project (Table 5.1), confirming the attention paid in the previous phase of pre-selection of the indicators, and having proposed extremely specific and priority indicators.

Although the choice of indicators always involves a subjective evaluation or, in any case, a link with the professional deformation of the discipline that a stakeholder represents, the selection process has shown that it has implemented a scientific and robust approach and, with the wide participation obtained, led to a balanced result

Table 5.1 Key performance indicators (KPIs) at the neighborhood scale

Indicator	Unit of measure
Availability of urban green areas	[sqm/inhabitants]
Availability and proximity of services of general interest to residential buildings	[N ° of services/sq km]
Incidence of urban green on the waterproof surface	[sqm of urban green/sqm waterproof surface]
Intermodality of the urban transport system	[N ° of intermodal nodes / sq km]
Residential population density	[N° of residents/sq km]
Mental health indicator (SF36)	[Standardized average scores]
Consistency of the network of cycle paths	[sqm of cycle area/sqm of land area]
Time devoted to mobility	[Survey: N° of minutes / 24 h]
Territorial density index	[land area/territorial area]
Total average annual consumption of thermal energy for the operation of residential buildings	[kWh/sqm/year]
Total average annual consumption of electric energy for the operation of residential buildings	[kWh/sqm/year]
Greenhouse gas emissions from the energy used for the operation of residential buildings	[kg CO_2 eq./sqm/year]
Density of pedestrian traffic areas	[sqm of sup. Pedestrian/sqm land area]
Waste: production of municipal waste	[kg/inhabitants]
Domestic water consumption	[liters per inhabitant per day]

inherent in the territorial context under study (Pignatelli 2020). Below is the list of the 15 validated KPIs:

5.2.2 *Phase 2: A Baseline Scenario*

The validation of the KPIs (Fig. 5.2) and the related calculation methodologies, units of measurement, and tools introduced the project to the next phase: the operational phase that is concerned their mapping and numerical quantification.

The objective was to obtain the basic data and materials for the calculation of the selected post-COVID indicators, data that had to be updated, corrected, extensible to the entire municipal area georeferenced as much as possible.

The results obtained from the spatialization of the first 6 KPIs (Table 5.1) for which the relative technical data sheets containing the calculation procedures have been developed are shown below. The technical sheets produced for each selected KPI have been divided into *Intent*—explanation of the indicator and its specificity concerning

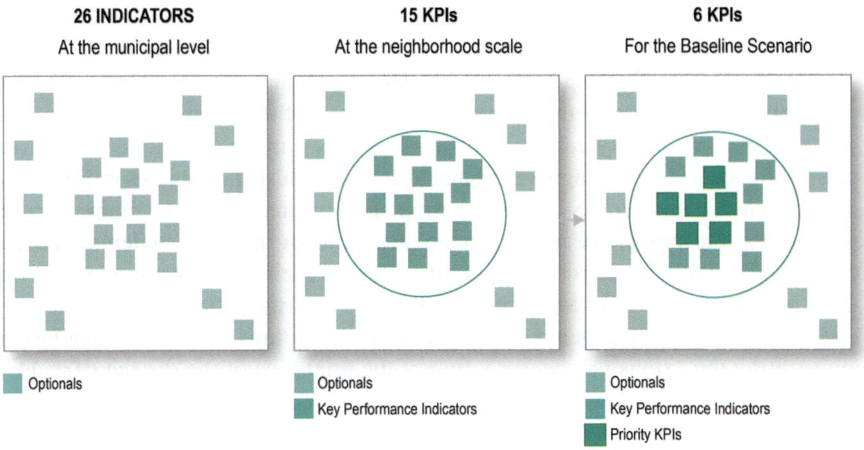

Fig. 5.2 Presentation of the final set of post-COVID indicators

the project; *Evaluation method* structured in Data requirements, Source of data, Evaluation method; and finally, the relative *Results*, maps showing the spatialized and calculated indicators.

The present indicators have been evaluated through spatial models and tools of the geographic information system (GIS) to map the results, and consequently, the method implements a new system based on multi-criteria indicators to support the decision-making process (Pignatelli 2020) and evaluate the neighborhood performance within the neighborhood.

This phase also concerned the application to the case study by experimenting with the methodology proposed within the "Crocetta" district of the Municipality of Turin. The choice of this district derives from the fact that it has been the seat of the Polytechnic of Turin since 1950 and from being a district in which there are important functions such as the Torino Porta Nuova railway station, the presence of three underground stations, and the Mauriziano Hospital, one of the main hospitals in the city.

5.3 Results

From the spatialization, evaluation, and analysis of the current state of the neighborhood, it has emerged that the "Crocetta" neighborhood represents one of the best-equipped neighborhoods in terms of services within its municipal area, in particular the small and large distribution, internal school services, and health services were analyzed. Indeed, from the analysis of their availability and proximity, it has emerged that all the residential buildings in the neighborhood are supplied at a walking distance of about 10/15 min (Balletto et al. 2020).

From the cross-reading between the availability of services and the number of transport modes present within it, the widespread distribution of bus stops, as well as other faster forms of transport such as underground and train, was demonstrated, allowing for a connection of the district with all parts of the city.

The presence of the Polytechnic of Turin plays a very important role for the district in promoting and encouraging its students to travel with sustainable mobility, since it can be seen that among the 13 Bike-sharing stations present for the entire district, 5 are close to the university as a means of travel most used by students who are attentive and sensitive to these issues, with 11 km of cycle paths available.

The neighborhood shows a scarce availability of internal urban green areas. Only 8% of the total area is covered by green areas (0.25 km^2 out of 3.4 available), demonstrating the extensive waterproofing of the total area of the district (92%).

5.4 Conclusion and Future Developments

In this chapter, a proposed methodological framework has been proposed in order to investigate the correlations between the urban environment (Pirni et al. 2020) and the level of proximity of the cities for the Basic Scenario of the city of Turin at the neighborhood scale, providing objective results on which future actions can be based.

For what has been said, a city that promotes proximity in the urban context in balance is farsighted, capable of planning and building its future by adapting to the unpredictable emerging challenges, such as the recent health emergency, thus being able to sustain, regenerate and re-establish a new balance within the urban environment. An effective assessment of complexity, differentiated for each territorial system, is necessary to develop and implement successful strategies to achieve common objectives. For this reason, multi-criteria analysis tools, analysis of qualitative and quantitative indicators, and performance evaluations able to measure the effectiveness of actions before their implementation are becoming increasingly important (Genta et al. 2019).

Based on the analysis of the post-COVID indicators proposed in this study, in particular the evaluation of the Key Performance Indicators, it was possible to highlight the weak points and priority areas on which to act, experimenting with the theme of "experiencing proximity to the neighborhood scale" as an urban response to the pandemic.

Measuring, evaluating, and spatially visualizing the impact were the keys to the proposed evaluation process, and the methodology resulted:

Replicable: The proposed methodology allows it to be replicated in all urban contexts.

Adaptable: An adequate methodology has been provided to create one's own set of indicators on which to base the analyses according to the topic being studied.

Accurate: The analyses were conducted at the neighborhood scale to be as accused as possible.

Specification: The proposed methodology allows to identify of the key indicators of the project (KPI).

The following are the main recommendations for future developments and implementations:

Evaluation and spatialization of the entire final set: This work experimented with the methodology proposed at the neighborhood level, developing the first 6 most priority KPIs one year after the start of the project. It is hoped that the methodology will then be applied to the entire municipal area by the Post Un-lock Project to obtain a general and broader picture of the situation.

Development of future scenarios: An extremely interesting aspect, which would further enrich this work, is the development of future scenarios. This would allow the comparison of urban assets before and after the modification, thus being able to compare the situations across different time-frames.

In this context, the methodology proposed in this study can provide the appropriate starting point for future research works based on the analysis of indicators and the use of multi-criteria spatial decision support systems, thus confirming the initial objectives set.

References

Aditi R, Johns P (2020) Rethinking cities in a post-COVID-19 World. Global Policy J

Angelakoglou K et al (2019) A methodological framework for the selection of key performance indicators to assess smart city solutions. Smart Cities 2(2):269–306. https://doi.org/10.3390/smartcities2020018

Balletto G et al (2020) A methodological approach on disused public properties in the 15-minute city perspective. Sustainability (switzerland) 13(2):1–19. https://doi.org/10.3390/su13020593

Capolongo S et al (2020) COVID-19 and cities: from urban health strategies to the pandemic challenges. A Decalogue of Public Health opportunities

Chelleri L (2012) From the "resilient city" to urban resilience. a review essay on understanding and integrating the resilience perspective for urban systems. Documents D'analisi Geografica 58(2):287–306. https://doi.org/10.5565/rev/dag.175

Chelleri L et al (2012) Multidisciplinary perspectives on urban resilience. In: Workshop Report, 1st edn. Basque Centre for Climate Change. Available at: http://resilience2014.sciencesconf.org/file/78005

Genta C et al (2019) Key Performance indicators for sustainable urban development: case study approach. IOP Conf Ser: Earth Environ Sci 296(1). https://doi.org/10.1088/1755-1315/296/1/012009

Pignatelli M (2020) Innovare i processi di pianificazione territoriale indicatori e scenari per valutare lo sviluppo urbano sostenibile: il caso studio del comune di torino

Pirni A, Caporale C (2020) Pandemia e resilienza. Persona, comunità e modelli di sviluppo dopo la Covid-19. https://doi.org/10.48220/PANDEMIAERESILIENZA-2020

Torabi Moghadam S, Lombardi P (2018) Building and environment. An interactive multi-criteria spatial decision support system for energy retrofitting of building stocks using Communtiy VIZ to support urban energy planning. Build Environ

Chapter 6
Towards Neighbourhoods as Minimum Units of Resilience?

Elena Pede, Mattia Scalas, and Luca Staricco

Abstract COVID-19 restrictions have changed the perception of space: travel limitations and diffusion of remote activities have narrowed the spaces of everyday life, leading to a rediscovery of proximity. We are both witnessing a re-appropriation of the domestic environment and re-discovering the neighbourhood and those small portions of the city often neglected. This rediscovery is evident in the use of nearby public spaces and in transport, with the decongestion of many urban areas following the reduction of commuting. Even if this is a contingent situation, it is reasonable to think that part of these changes will persist at the end of the emergency. For these reasons, there is a need to focus on neighbourhoods' quality, spatial organization and adaptive capacity towards both emergencies such as the pandemic and the great urban challenges towards resilience and sustainability. Essentially, sub-local scale must be rethought to meet not only the ordinary needs of its inhabitants but also health or other issues. In this sense, the potential of spatial units based in the concepts of proximity and walkability is explored, giving an interpretation that starting from the 15-min city and the superilla models explore the perspective of "minimum units of resilience" for facing pandemics.

Keyword COVID-19 · Resilience · Sustainability · Neighbourhood planning · 15-min city · Superilla

E. Pede (✉) · M. Scalas · L. Staricco
DIST—Interuniversity Department of Regional and Urban Studies and Planning,
R3C—Interdepartmental Responsible Risk Resilience Centre, Politecnico di Torino, Turin, Italy
e-mail: elena.pede@polito.it

M. Scalas
e-mail: mattia.scalas@polito.it

L. Staricco
e-mail: luca.staricco@polito.it

G. Brunetta et al. (eds.), *Post Un-Lock*, The Urban Book Series,
https://doi.org/10.1007/978-3-031-33894-6_6

6.1 Neighbourhoods and Pandemic

Urban areas have always been exposed to pandemics because of high density, mass interactions, connections and concentration of activities. Since ancient time, the emergence of a pandemic has posed the attention to cities as hotspots for the spread of diseases. Each pandemic has seen the re-thinking of urban form, sanitation systems, street design and housing regulations together with rules, ordinances and restriction to reduce the threat of contagion.

During the Industrial Revolution, as a result of the disease outbreaks linked to water and unhealthy air, there was an overwhelming demand for the improvement of urban environments to reduce the threat of disease. Consequently, public health interventions expanded to encompass issues like housing, sewerage, clean water provision and nutrition. In the same way, many cities have imposed restrictions, mask wearing and disinfection of shared spaces due to the Spanish flu.

Despite the progress during the past century, the pandemic caused by COVID-19 still caught cities unprepared. Although technological innovations and new urban form paradigms, the pandemic has transformed almost every aspect of contemporary urban living.

During the past two years, short-term interventions have been implemented around the world. Work and lessons from home were boosted thanks to online resources, and many countries have implemented restrictions on mobility in order to prevent the spread of Coronavirus disease-19, disrupting daily mobility. Once again, the consequences of pandemic and the need to coexist with the emergency have made clear that a paradigm shift is required.

Part of the debate has focused on the performance of cities in relation to contagion levels and mortality rates, but there has also been significant discussion around the role that the urban form and function, spatial patterns and access to urban services may play in living this new normality (UN-Habitat 2021). If in the early months of the pandemic the challenge was indeed to flatten the contagion curve at the cost of putting the world on an unprecedented pause; in the second wave of the virus—also because of pressing socio-economic issues—governments and cities needed to preserve quality of life despite the uncertainty.

In a context where social distancing was suddenly essential, issues like proximity, local neighbourhood, public space, accessibility to essential urban services and different transport modes have captured the attention of many politicians and have prompted cities to work on it.

Travel within cities has been strongly limited with lockdowns and restrictions of out-of-home activities. At the same time, people also applied self-restrictions avoiding public transport in favour of private mobility (car and active travel modes such as walking and cycling) due to their perception of risk of virus transmission on the different modes (Hanzl 2020). Cities witnessed the huge increase of cycling and pedestrian mobility demand and have re-designed their streets to accommodate bikers and pedestrians needs (Moreno et al. 2021).

Similarly, the long-term consequences of restrictions on the use of public space, and the access to urban services has also transformed our relationship with open spaces and proximity. Green spaces were considered to be increasingly important as well as public squares and streets; they provided space for physical, social and cultural activities with a lower risk of infection. For the same reasons, private or communal outdoor spaces have experienced a renaissance in the quality of life perception in the cities. It is not by chance that disparities in housing conditions and accessibility to facilities and green areas have magnified the inequalities between the residents of different urban neighbourhoods.

The role of public open spaces was crucial also for physical and mental health and for face-to-face interactions. While in business districts cafes, restaurants, shops and cultural spaces have been strongly hit by work-from-home habits, in residential neighbourhoods these activities—especially those with outdoor areas—have contributed to reactivate local community and social interaction (Mouratidis 2021). Dwellers needed to find social, economic and cultural activities in their immediate localities in order to reduce travel, and this was an opportunity to rediscover neighbourhoods' shops. The neighbourhood unit and the social ties within it have strongly been reshaped, and it is supposed that many of these changes will have long-term consequences in urban form and planning.

While the pandemic has transformed almost every aspect of urban living, new narratives about urban form, access to services and amenities have emerged. Planning models like *superilles* or *15-min city* that were mainly born in the frame of sustainability goals offer attractive perspective for building safer, more resilient, sustainable and inclusive cities. Their organization of street networks as well as the distribution of services and amenities emphases the concepts of accessibility and proximity that are well suited to the challenges posed by COVID-19 pandemic. The convergence of different aims suggests long-term implications for urban resilience. Through this approach, the chapter explores the potential of spatial units based on walkability as the minimum system capable of reacting to crises effects of different nature. The *superilla* and the 15-min city models are introduced (as re-interpretations of the XXth century's neighbourhood unit) and analysed just in their perspective role of minimum units of resilience for facing pandemic crises.

6.2 The Ideal Type of the Neighbourhood Unit

The neighbourhood unit has been acknowledged as one of the main landmarks in the last century urban planning (Patricios 2002). It can be briefly described as a spatial model which aims to be self-sufficient in terms of basic service provision and to ensure safe accessibility to these services by walking and cycling (Rohe 2009).

Even if the expression was first coined by William Drummond in 1916, the neighbourhood unit is usually traced back to the contribution of Clarence Perry to the first *Regional plan of New York and its environs* in 1929 and to Clarence Stein and Henry Wright's project of Radburn, the "town for the motor age", in the same year

(Brody 2016). They both implement a traffic separation principle: T-intersections and cul-de-sac are adopted in the inner streets of the unit to discourage car-through traffic (which is diverted to the thoroughfares deliming the unit) and to allow only access traffic; separated footpaths are provided to move safely inside the unit and to access the basic services (*in primis* the primary school, but also the library, the church, the playground, etc.) (Grammenos et al. 2008). These services are primarily located in a barycentric position in the unit, with the exception of shops, which are concentrated in a boundary corner so to be accessible also to the other surrounding units.

Clearly inspired by the wards of Howard's Garden city, the neighbourhood unit has been "a persistent idea" in twentieth-century urban planning (Birch 1980), not only in America. Although not free of criticism [relating to social and functional segregation, as well as physical determinism; (Mehaffy et al. 2015)], this idea has been repeatedly proposed in different contexts and times, from second Post-war Abercrombie's Greater London Plan and British New Towns to Buchanan's Environmental areas, until recent proposals by the Congress for the New Urbanism, European eco-district experimentations and superblocks in the Middle East and China.

6.3 The Spanish *Superilla*

While most of the implementations of the neighbourhood unit model were related to new urban developments, in the last twenty years a new reinterpretation of the neighbourhood unit—the so-called superilla (or supermanzana, in Catalan)—was proposed in the Spanish context to be applied to well-established built areas. The superilla concept has been elaborated by Salvador Rueda, director of the Agencia de Ecologia Urbana de Barcelona (Urban Ecology Agency of Barcelona—UEAB), in the framework of his "Ecological urbanism" approach (Rueda 2014). It is an ideal 400 × 400 m "urban cell" which is conceived as a basic unit for reorganizing the existing city and increasing its sustainability by applying the traffic separation principle. Streets are hierarchized into two levels. Those inside the superilla are supposed to support only car traffic whose origins or destinations are inside the superilla itself; cut-through traffic is discouraged in these streets by maximum speed of 10 km/h, frequent changes of direction, traffic calming, etc. Instead, those streets that border the superilla are designed as thoroughfares to host cut-through traffic, public transport (buses, taxis, tramways, etc.) and bike lanes. By spreading this model on the whole city, a two-level new urban morphology is established: a main street network, which is made up of thoroughfares that cross every 400 m; and the meshes of this network, which are the superilles, whose inner streets are intended mainly for walking and for multiple local uses and functions.

The superilla model has been tested in a number of Spanish cities (such as Victoria-Gasteiz, Ferrol, Madrid and Valencia), but Barcelona is the one that has most experienced it. At the city scale, the *Pla de Mobilitat Urbana 2013–2018* identified the main network of thoroughfares, whose speed limit was set to 50 km/h; the meshes of this network were classified as superilles (mostly corresponding to a 400 × 400 m square made up of 3 × 3 blocks), whose inner streets had a 10 km/h speed limit. The *Pla* proposed the reorganization of the bus system, creating a "Nova Xarxa de Bus", made up of 28 bus lines riding along the thoroughfares, without crossing the superilles. In addition, over 200 km of new bicycle lanes were planned on the main network in order to guarantee a high-speed cycling mobility. At the neighbourhood level, the *Omplim de vidas els carres* programme adopted by the city administration in 2016 identified one pilot superilla in each of the ten city's districts to be realized. As some analyses of these pilot projects highlighted, this spatial model seems to be more easily accepted when applied to urban areas featuring a clear and physically evident hierarchy between main and secondary streets (such as Gracia and Ciutat Vella in Barcelona), than to homogeneous street networks such as the Cerdà grid (Scudellari et al. 2020; Zografos et al. 2020).

In the very last years, Barcelona's administration had changed its approach: instead of proceeding by areas, incrementally implementing the new superilles one after the other, the strategy was focused on creating a network of "green streets and squares" across the whole city, by enabling some streets to be freed of road traffic and to give priority to pedestrians. This approach is being tested on the whole district of the Eixample, where 21 streets (for a total length of 33 km) and 21 squares at the intersection of these streets (for a total surface area of 3.9 ha) should be re-designed so to allow cars to circulate at a maximum speed of 10 km/h and ensure priority for pedestrians and cyclists. The aim is to provide the Eixample with 33.4 ha of new pedestrian areas and 6.6 ha of urban green areas; in this way, each resident in the district will have a square or green street within 200 m of his homes (Staricco and Brovarone 2022).

6.4 The 15-Min City

The neighbourhood unit can be considered a clear reference also for the 15-min city model (especially for its 20-min neighbourhood variation in American and Australian low-density cities). This model first appears in Carlos Moreno's 2016 book "*Droit de la Cité—De la «ville-monde» à la «ville du quart d'heure»*", attracting attention and critics of scholars, policy makers and citizens. Sometimes defined as a *eutopia* (Pozoukidou and Chatziyiannaki 2021), the 15-min city is proposed as a model of spatial organization of cities based on pedestrian or cycle accessibility within a quarter of an hour to all the services of the daily needs of citizens. This chrono-urbanism perspective was often interpreted as a *manifesto* (Marchigiani 2021) for its clarity, simplicity and ability to hold together a multitude of different concepts and dimensions, spreading outside the scientific debate and being adopted as a city paradigm proposed by Anne Hidalgo during her 2020 campaign for the Paris local elections. The 15-min city aims to overcome a vision of the city focused on car dependence, a modernist approach—Moreno explicitly refers to Le Corbusier—which contributed to the development of the lattice structure of cities and to sprawl.

Developed as a model to reduce dependence on cars and therefore promote the reduction of emissions and sustainability, according to Moreno the 15-min city can also be seen as a paradigm for overcoming the legacy of car-dependent urban organization with its deep-rooted social and economic inequalities (Moreno et al. 2021). The attention on the 15-min city has grown further with the COVID-19 pandemic (Pinto and Akhavan 2022): the impact of restrictions and social distancing started a debate on density and the rediscovery of proximity. In this context, in 2021 Carlos Moreno revived the concept, defending the dense city and reinterpreting the 15-min approach as a planning response to the pandemic. In this sense, the pandemic proved to be a sort of case study, which highlighted the shortcomings of cities in terms of distribution and accessibility of services, but also allowed to visualize the effects that some of the most desired policies, such as traffic reduction and the increase in the use of bicycles, may have on settlements. The restrictive lockdown policies were in fact also accompanied by numerous temporary urban planning initiatives, especially related to accessibility of green areas, the establishment of new cycle lanes (Mexico City) or the closure to motor vehicles (Vienna, Boston) (Krzysztof and Drozda 2021). According to Moreno, the permanent adoption of these policies would lead to an increase in the quality of life in the cities enshrined in the reduced commuting times, proving the effectiveness of the 15-min city or similar declinations such as the 20-min neighbourhood (Gower and Grodach 2022). The most updated definition of the 15-min city model identifies six key functions to support a dignified urban life, identified in living, working, commerce, healthcare, education and entertainment, while there are four dimensions that should be incorporated in the planning it intends to pursue this paradigm: proximity, density, diversity and digitalization, the latter added in the re-elaboration of the concept in a post-pandemic key (Moreno et al. 2021).

Despite an objective difficulty in identifying an operational definition, the 15-min city has been recognized by numerous international organizations as a useful prospect for reaching SDG11 "sustainable cities and communities", with application examples not only in Paris but also in New York, Bogotà, Melbourne and the Milan "phase-2" post-pandemic strategy. C40 actively promote the dissemination of this planning paradigm, with publications, indications and guidelines addressed to local administrators, whose electoral campaigns that often have taken up Hidalgo's approach proposing the model as a vision for the city. At an operational level, the 15-city concept has often been interpreted in terms of walkability analysis and accessibility to services, leading to different interpretations of what these services should be. Models have been developed mainly focused on the construction of the network and the calculation of isochronous curves, for example with respect to schools within 15 min on foot from the maximum concentration of population in a neighbourhood (Caselli et al. 2022). There are also attempts to calculate overall indexes of the closeness of a territory to the model of the 15-min city, arising from interactions between private sector and academia to promote city smartness. An example is the 15-min index made available by the energy company Enel X to Italian local administrations and developed with the University of Florence (Badii et al. 2021). This tool, fully based on open data, develops 13 indicators starting from the six themes proposed by Moreno and calculates them dividing the national territory into squares each one representing accessibility in 15 min (Nesi and Gambacorta 2021). The result is an index that evaluates the territory readiness with respect to the 15-min model, helping administrations in policy design. However, these kinds of initiatives show that the 15-min city model still lacks operational definitions and needs specific interpretations to be applied. Furthermore, the model may not be sufficient to identify planning priorities in territories that have already introduced in their regulations a distribution of minimum services, even if built based on the number of settled inhabitants.

6.5 Are Minimum Units of Resilience a Worthwhile Target?

The COVID-19 pandemic has highlighted the important role of urban liveability at the neighbourhood scale during a health emergency. On the one side, in the last months, medium and long distance trips have become more difficult, due to a couple of reasons: public transport services have been reduced or suspended in most cities, as overcrowded buses, trams, tube, etc., were pointed out as a high infection risk factor (Das et al. 2021); stringent health protocols, social distancing, lockdowns and movement restrictions have been adopted to face the pandemic. That made evident the necessity of a minimum set of proximity-based services that should be accessible by walking or cycling in the short distance (Marin-Cots and Palomares-Pastor 2020). On the other side, public spaces (such as green areas, playgrounds, courtyards, etc.) turned out to be essential to perform outdoors some activities that could no longer be carried out indoors in schools, gyms and so on.

In other words, a neighbourhood offering: (1) a fundamental set of basic services, (2) a network of foot- and cycle-paths to easily and safely reach those services by walking and cycling and (3) public spaces for meeting and performing social activities outdoors seems to ensure a certain level of resilience, at least for facing pandemic risks. Therefore, does it make sense to conceive and design neighbourhoods as "minimum units of resilience", at least in relation to pandemic risks? And can the city be planned as a "mosaic" of these resilience units?

The neighbourhood units launched about a century ago by Perry, Stein and Wright in the USA largely fulfilled the three above-mentioned requirements. Therefore, it is probably no coincidence that the superilles and the 15-min city, two spatial models clearly inspired by the neighbourhood units, have gathered great momentum and have been embraced in a growing number of scholars and city mayors' agendas during the pandemic. At the same time, both these models have a number of critical issues that question their actual resilience.

As regards the superilles, they are focused on reducing the street section devoted to car circulation and parking in order to free up public space for walking and cycling, and for performing outdoor activities. Little or no attention is paid to offering a set of basic services inside the superilla, also because its average dimension is too limited for ensuring the minimum number of customers and users that is necessary for the efficiency of a primary school, a library, etc. Only in the recent evolution of this model, focused on creating a network of "green streets and squares" in the

Eixample district, there is a proximity target: each resident in the district should have a square or green street within 200 m of his home. Moreover, as one of the first implemented superilla (the one in Poblenou) has shown, this model can generate a number of conflicts and discriminations, between residents inside the superilla and residents in surrounding neighbourhoods; residents and activities inside a superilla and those located at its verges; residents and visitors of the superilla; car drivers and non-motorized citizens; pedestrians and cyclists.

Also the 15-min city, and particularly its 20-min neighbourhood variation adopted in low-density cities, cannot be considered strictly resilient. Paradoxically, till now this concept has often been at risk of being reduced to a mere political slogan (Andres Duany 2021). In many cases, it was translated in a query for traditional issues such as density, diversity and mix of land uses, walkability, etc.; instead, poor attention was paid to the effective accessibility by proximity of the services in the neighbourhood or the city: which services should be accessible within 15 min? Which is the indivisibility threshold (i.e. the minimum number of users) for these services? Why was the temporal threshold set precisely at 15 min? These and many other issues, essential for operationalizing the 15-min city concept, often remain unanswered in practices.

Beyond these specific weaknesses of the two models analysed in this paper, further fundamental problems related to the neighbourhood unit approach can re-emerge when implementing resilience at the neighbourhood scale. These problems mainly concern the relation between the unit and the rest of the city. On the one side, neighbourhood (or resilience) units are at risk of creating social and spatial segregation and gentrification (Mehaffy et al. 2015). On the other side, diverting the cut-through traffic to the borders of these unit can determine what Jacob (1961) over 50 years ago termed "the curse of border vacuums".

In conclusion, all these critical issues should warn against adopting the neighbourhood—through an excessively simplifying approach—as the right spatial scale to implement minimum resilience units. At least in densely populated cities such as the European one, it is not obvious that resilience could be pursued more effectively at this level and not at the whole urban scale, particularly if multiple risks—and not only health ones—are taken into account.

References

Andres Duany RS (2021) Defining the 15-minute city, CNU. Available at: https://www.cnu.org/publicsquare/2021/02/08/defining-15-minute-city. Accessed 12 Nov 2021

Badii C et al (2021) Computing 15MinCityIndexes on the basis of open data and services. In: 21st International conference computational science and its applications—ICCSA 2021, pp 565–579

Birch EL (1980) Radburn and the American planning movement the persistence of an idea. J Am Plann Assoc 46(4):424–439. https://doi.org/10.1080/01944368008977075

Brody J (2016) How ideas work: memes and institutional material in the first 100 years of the neighborhood unit. J Urbanism: Int Res Placemaking Urban Sustain 9(4):329–352. https://doi.org/10.1080/17549175.2015.1074602

Caselli B et al (2022) Exploring the 15-minute neighbourhoods. An evaluation based on the walkability performance to public facilities. In: XXV International conference living and walking in cities—new scenarios for safe mobility in urban areas (LWC 2021), Elsevier (Transportation Research Procedia), pp 346–353

Das S et al (2021) Impact of COVID-19: a radical modal shift from public to private transport mode. Transp Policy 109:1–11. https://doi.org/10.1016/j.tranpol.2021.05.005

Gower A, Grodach C (2022) Planning innovation or city branding? Exploring how cities operationalise the 20-minute neighbourhood concept. Urban Policy Res [Preprint]

Grammenos F et al (2008) Hippodamus rides to Radburn: a new model for the 21st century. J Urban Des 13(2):163–176. https://doi.org/10.1080/13574800801965643

Hanzl M (2020) Urban forms and green infrastructure–the implications for public health during the COVID-19 pandemic. Cities Health, pp 1–5

Jacob J (1961) The death and life of the American city. Vintage, New York

Krzysztof H, Drozda Ł (2021) Green infrastructure in the time of social distancing: urban policy and the tactical pandemic urbanism. Sustainability 13(4)

Marchigiani E (2021) Il progetto della "città dei 15 minuti": esercizi critici di prospettiva e di memoria. Urbanistica Informazioni, pp 7–11

Marin-Cots P, Palomares-Pastor M (2020) In a 15 minute setting. Towards the city of proximity, and its relationship with the covid-19 and the climate crisis: the case of Malaga. Ciudad Territorio Estud Territoriales, pp 685–700

Mehaffy MW, Porta S, Romice O (2015) The "neighborhood unit" on trial: a case study in the impacts of urban morphology. J Urbanism: Int Res Placemaking Urban Sustain 8(2):199–217. https://doi.org/10.1080/17549175.2014.908786

Moreno C et al (2021) Introducing the "15-minute city": sustainability, resilience and place identity in future post-pandemic cities. Smart Cities, pp 93–111

Mouratidis K (2021) How COVID-19 reshaped quality of life in cities: a synthesis and implications for urban planning. Land Use Policy 111:105772

Nesi P, Gambacorta S (2021) Webinar Enel X—Città a 15 minuti: come pianificare lo sviluppo urbano di prossimità sfruttando gli Open Data?

Patricios NN (2002) The neighborhood concept: a retrospective of physical design and social interaction. J Architectural Plan Res 70–90

Pinto F, Akhavan M (2022) Scenarios for a post-pandemic city: urban planning strategies and challenges of making "Milan 15-minutes city". Transp Res Procedia 60

Pozoukidou G, Chatziyiannaki Z (2021) 15-minute city: decomposing the new urban planning Eutopia. Sustainability 13(2):928. https://doi.org/10.3390/su13020928

Rohe WM (2009) From local to global: one hundred years of neighborhood planning. J Am Plann Assoc 75(2):209–230. https://doi.org/10.1080/01944360902751077

Rueda S (2014) Ecological urbanism: its application to the design of an eco-neighbourhood in Figueres. Agència d'Ecologia Urbana de Barcelona, Barcelona

Scudellari J, Staricco L, Vitale Brovarone E (2020) Implementing the Supermanzana approach in Barcelona. Critical issues at local and urban level. J Urban Des 25(6):675–696. https://doi.org/10.1080/13574809.2019.1625706

Staricco L, Brovarone EV (2022) Livable neighborhoods for sustainable cities: insights from Barcelona. Transp Res Procedia 60:354–361. https://doi.org/10.1016/j.trpro.2021.12.046

UN-Habitat (2021) Cities and pandemics: towards a more just, green and healthy future. United Nations Human Settlements Programme

Zografos C et al (2020) The everyday politics of urban transformational adaptation: struggles for authority and the Barcelona superblock project. Cities 99:102613. https://doi.org/10.1016/j.cities.2020.102613

Part III
Case Studies

Chapter 7
NO_2 Concentrations and COVID-19 in Local Systems of Northwest Italy

Ettore Sarzotti, Gianmarco Pignocchino, Alessandro Pezzoli, and Angelo Besana

Abstract This paper aims to further investigate the relationship between the concentrations of nitrogen dioxide (NO_2) and the severity of COVID-19 by analyzing the data of three Italian Regions (Piedmont, Valle d'Aosta and Liguria) during the first wave of the pandemic (February–May 2020). The analyses were conducted at a local scale using Local Labor Systems of ISTAT. The annual average of NO_2 concentrations, obtained from space satellite Sentinel-5P, was used to assess environmental data. While excess mortality data were used to estimate the severity of the pandemic, calculated as the percentage change in deaths recorded in 2020 compared to the average number of deaths of the previous five years (2015–2019). Using quasi-Poisson multivariate regression models, it was possible to estimate the correlation between the incidence rate of the pandemic and some risk factors, including in particular the concentration of NO_2.

Keywords COVID-19 · Nitrogen dioxide · Local systems

7.1 Introduction

The northern regions have been the most affected area by the first wave of the COVID-19 epidemic in Italy. The great speed and intensity with which disease has spread to these regions has led to the hypothesis, in some preliminary studies, that high levels of pollution may play a role in epidemic expansion and in determining the severity of the infection (Coccia 2020; Conticini et al. 2020; Fattorini and Regoli 2020; Martelletti

E. Sarzotti (✉) · G. Pignocchino
Università degli Studi di Torino, Turin, Italy
e-mail: ettore.sarzotti@gmail.com

A. Pezzoli
DIST—Interuniversity Department of Regional and Urban Studies and Planning, Politecnico di Torino, Turin, Italy

A. Besana
Humanities Department, University of Trento, Trento, Italy

G. Brunetta et al. (eds.), *Post Un-Lock*, The Urban Book Series,
https://doi.org/10.1007/978-3-031-33894-6_7

and Martelletti 2020). In fact, Northern Italy is one of the most heavily polluted area in Europe in terms of smog and air pollution (Carugno et al. 2016) because it is characterized by a high concentration of densely populated urban areas, as well as by a strong presence of industrial activities. In addition, the particular geomorphological conformation and the weather-climatic characteristics of some areas do not favor the recirculation and release of pollutants with their consequent stagnation.

This contribution intends to verify these hypotheses at a more detailed geographical scale than the previous studies (Fattorini and Regoli 2020; Pignocchino et al. 2022), that is to say at the level of local systems, considering the northwestern regions of Italy, that are characterized by different environmental conditions and socio-demographic patterns. The results should therefore have a greater geographical consistency and, therefore, scientific validity.

7.2 Materials and Methods

This research project focused on the period of the first COVID-19 epidemic wave (February–May 2020) in the Northwest regions. We worked on the data at the local level by considering Local Labor Systems (LLS), which are inter-municipal territorial units, independent from any administrative subdivision, and defined by the daily commuting flows recorded in the general population and housing censuses (ISTAT 2015a).

In order to analyze this data at the local level, the first necessary step was to create a cartography of LLS updated to the most recent changes in the administrative boundaries of the municipalities using the data referring to January 1, 2020, obtained by ISTAT—Italian National Institute of Statistics (ISTAT 2020a). The total number of LLS in Piedmont, Valle d'Aosta and Liguria regions is 55 (Fig. 7.1).

From the first months of the pandemic, a discrepancy is emerged between official data on COVID-19 deaths and the significantly higher number of deaths compared to previous years (Ciminelli and Garcia-Mandicó 2020). Therefore, to estimate the real impact of COVID-19, we used excess mortality data, defined as the percentage of change in the mortality of the year 2020 compared to the average number of deaths in the same period of the previous five years (2015–2019). These data were retrieved from ISTAT (ISTAT 2021a), temporally aggregated bi-weekly and then for the overall period (February-May).

Nitrogen dioxide (NO_2) concentrations data were derived from the space satellite Sentinel 5 Precursor (S5P) managed by the European Space Agency (ESA) under the Copernicus program. According to Eum et al. (2019), there is an association between one year of exposure to nitrogen dioxide and the increase of mortality from several pathologies (cardiovascular and respiratory, in particular). Therefore, we decided to consider the time period of one year as proxy for the pollution data. The data were retrieved from the Sentinel-5P OFFL NO_2 dataset through Google Earth Engine platform (GEE), obtaining a single grid image defined by the annual mean of NO_2 concentrations from July 1, 2018 to June 30, 2019 with a resolution of 1 km^2 per

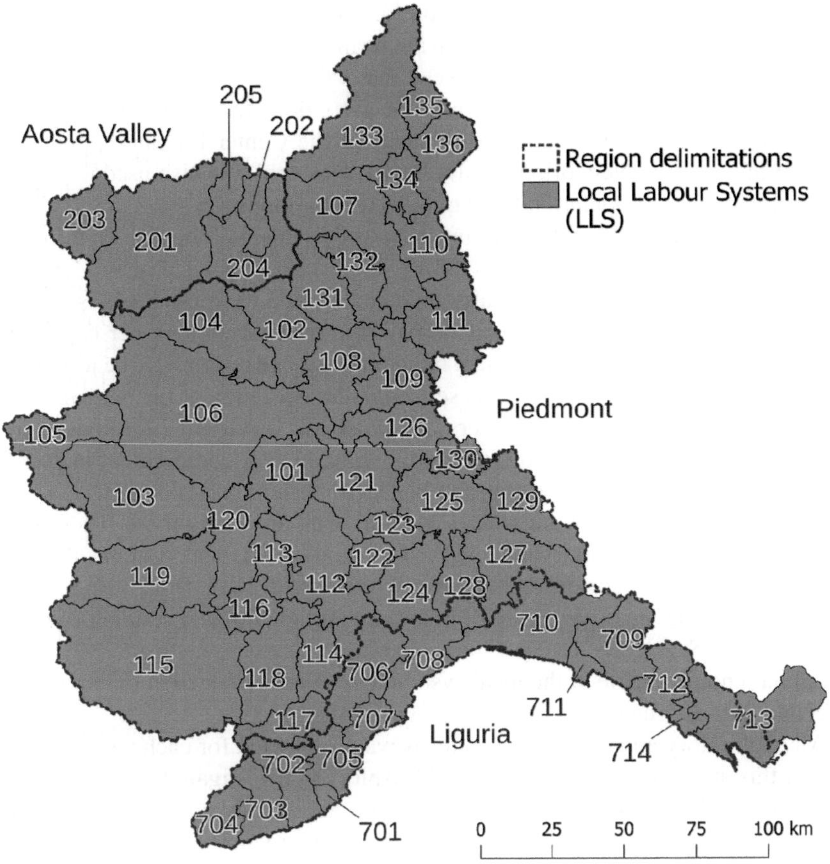

LLS Code	LLS Denomination	LLS Code	LLS Denomination	LLS Code	LLS Denomination
101	CHIERI	120	SAVIGLIANO	201	AOSTA
102	IVREA	121	ASTI	202	AYAS
103	PINEROLO	122	CANELLI	203	COURMAYEUR
104	RIVAROLO CANAVESE	123	NIZZA MONFERRATO	204	SAINT-VINCENT
105	SUSA	124	ACQUI TERME	205	VALTOURNENCHE
106	TORINO	125	ALESSANDRIA	701	DIANO MARINA
107	BORGOSESIA	126	CASALE MONFERRATO	702	IMPERIA
108	SANTHIÀ	127	NOVI LIGURE	703	SANREMO
109	VERCELLI	128	OVADA	704	VENTIMIGLIA
110	BORGOMANERO	129	TORTONA	705	ALBENGA
111	NOVARA	130	VALENZA	706	CAIRO MONTENOTTE
112	ALBA	131	BIELLA	707	FINALE LIGURE
113	BRA	132	COSSATO	708	SAVONA
114	CEVA	133	DOMODOSSOLA	709	CHIAVARI
115	CUNEO	134	OMEGNA	710	GENOVA
116	FOSSANO	135	SANTA MARIA MAGGIORE	711	RAPALLO
117	GARESSIO	136	VERBANIA	712	SESTRI LEVANTE
118	MONDOVÌ			713	LA SPEZIA
119	SALUZZO			714	LEVANTO

Fig. 7.1 LLS of Piedmont, Valle d'Aosta, and Liguria

pixel.[1] Pollution data of nitrogen dioxide for the first period of the epidemic wave (February to May) were also collected in the form of nine grids, each with the average concentration referenced over a two-week interval.

Next to this, population-weighted average was calculated using a grid image from the Gridded Population of the World dataset of the Center for International Earth Science Information Network (CIESIN 2018). This grid was also used to derive the population density (population/km^2) of each local system.

The Italian National Institute of Statistic was the source of the data on the resident population as of January 1, 2020 (ISTAT 2020b) and for LLS quality indicators (ISTAT 2015b). Among many quality indicators, three were used in this study: the number of jobs, from which it was derived the number of jobs per capita (jobs/population), and two self-containment indices (SCI), one referred to the jobs demand and the other to the jobs supply. The self-containment index of the demand (SCID) is calculated as the number of workers who work and live in the LLS divided by the number of workers who are working there, while the self-containment index of the supply (SCIS) is calculated as the rate between the number of workers who work and live in the LLS and the number of workers who are living there. These indices vary from 0 to 1 and quantify the level of isolation of the local systems, i.e., higher values of SCID index correspond to fewer workers who come from outside the LLS to work, while higher values of SCIS index correspond to fewer residents who work outside the LLS.

All data not available at the local system level were obtained at the municipality level and then aggregated to LLS.

An exploratory and descriptive analysis was conducted for each variable considered in this study, using histograms and boxplots to investigate the distributions of the values and detect the presence of outliers, and finally cartograms to evaluate the spatial distribution. To assess the effect of emergency measure taken by the Italian government (lockdown) on air pollution, a series of maps of the bi-weekly period of NO_2 concentrations was created. In addition, for each bi-weekly period, a cartogram on excess mortality data was also created to visualize the spatial and temporal evolution of the pandemic.

Spearman's correlation coefficient[2] was used to examine the relationship between nitrogen dioxide concentrations and excess mortality. Next to this, three quasi-Poisson regression models were implemented to exploring further the relationship between the two variables. Poisson's multivariate regression models are generalized linear models based on a logarithmic scale. They are useful with counting variables (number of deaths, in this case) and for modeling rates. A generalization of Poisson models was used in this study: the quasi-Poisson regression, which allows to take

[1] The tropospheric vertical column of NO_2 was considered and the concentrations values were transformed from mol/m^2 to μmol/m^2 for an easier consultation.

[2] A nonparametric index based on the ordinal position of the values (ranks), which fits better with variables that present outliers since is less affected from them. This coefficient varies from -1 to $+1$, where values close to zero indicates no correlation and values close to 1 means a strong correlation (negative or positive).

in account the overdispersion of the data by adjusting the variance to a specific dispersion parameter (Berk and MacDonald 2008).

Within these models, other variables were considered for a possible confounding effect on the associations between the two main variables: the population density, the proportion of the population over the age of 65 (population over 65/total population), the proportion of the male population (male population/total population), the number of jobs per capita, the LLS self-containment index of the demand, and the LLS self-containment index of the supply. These confounding factors were chosen as the data show higher mortality among males and among the elderly people. More deaths are expected in areas with higher population density while higher number of jobs and the level of isolation measured by the self-containment indices could represent a limitation to the spread of the virus within local systems.

Results obtained from the quasi-Poisson multivariate regression models are expressed as estimated rate ratio (RR), a relative difference used to compare different observed incidence rates due to exposure to a given risk factor over a given time period. In this study, the rate ratio is calculated as the ratio of the incidence rate in an exposed group and the incidence rate in a less exposed group. The risk factor is the NO_2 concentrations.

All cartograms were created with QGIS platform and all statistical analysis were carried out with RStudio software.

7.3 Results

7.3.1 Descriptive Analysis

All the variables used in this study have a tendency toward a normal distribution, as shown in the histograms in Fig. 7.2 The most extreme values, numerically distant from the rest of the data collected, are identifiable as outliers and only the two self-containment indices are the variables that do not present with one, as also confirmed by the boxplots in Fig. 7.3.

The effects of the limitations to circulation, imposed by the Italian government with a national lockdown, in LSS considered are visible in Fig. 7.4. The reduction in NO_2 concentrations starts from the 9th week, and there is a constant decreasing trend of the values recorded for the whole lockdown period (weeks 11–18). The end of the national restrictions marks the return to higher values as shown in weeks 19–20.

A good visualization of the first evolution wave is represented in Fig. 7.5. The cartograms show the excess mortality in the nine bi-weekly periods considered from weeks 5 and 6 (January 27–February 9) to weeks 21 and 22 (May 18–31) in all LLS used in this study. It is evident that the excess mortality started in weeks 9–10 when the number of local systems with excess mortality was greater than 50% (Fig. 7.6). The peak was reached in weeks 13–14, with only one local system without excess mortality, and began to fade after weeks 15–16. It is noticeable that the most affected

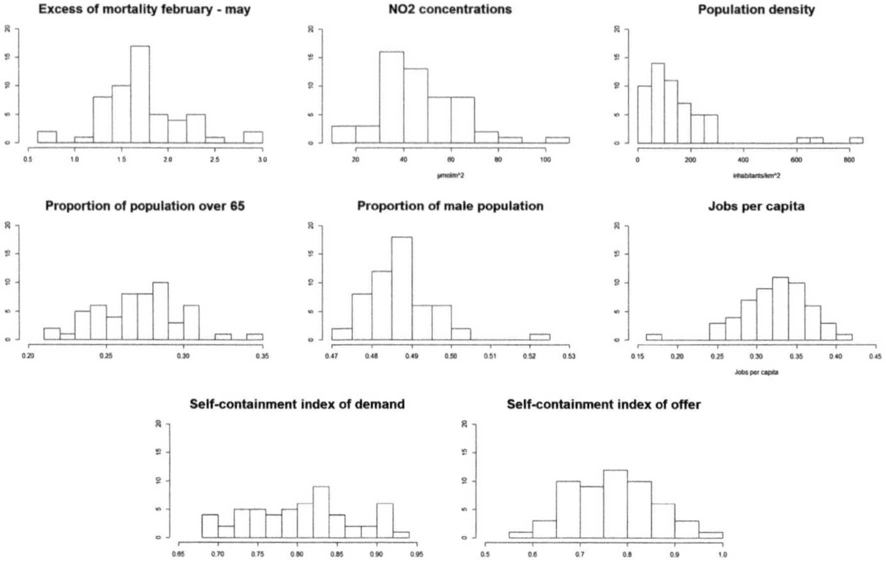

Fig. 7.2 Histograms of variables

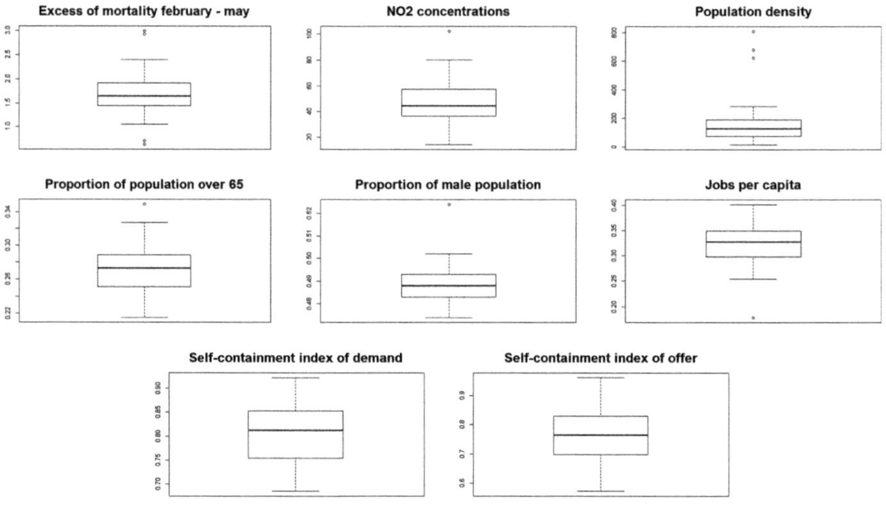

Fig. 7.3 Boxplots of variables

period, which runs from weeks 11 to 18, corresponded to the national lockdown that started on March 9 (DPCM 2020a) and ended on May 4 (DPCM 2020b).

Considering the excess mortality for the whole period of the first wave (Fig. 7.7a), a pattern of higher values emerges close to the border with Lombardy region, although not well defined since some border LLS report only medium values. Other clusters

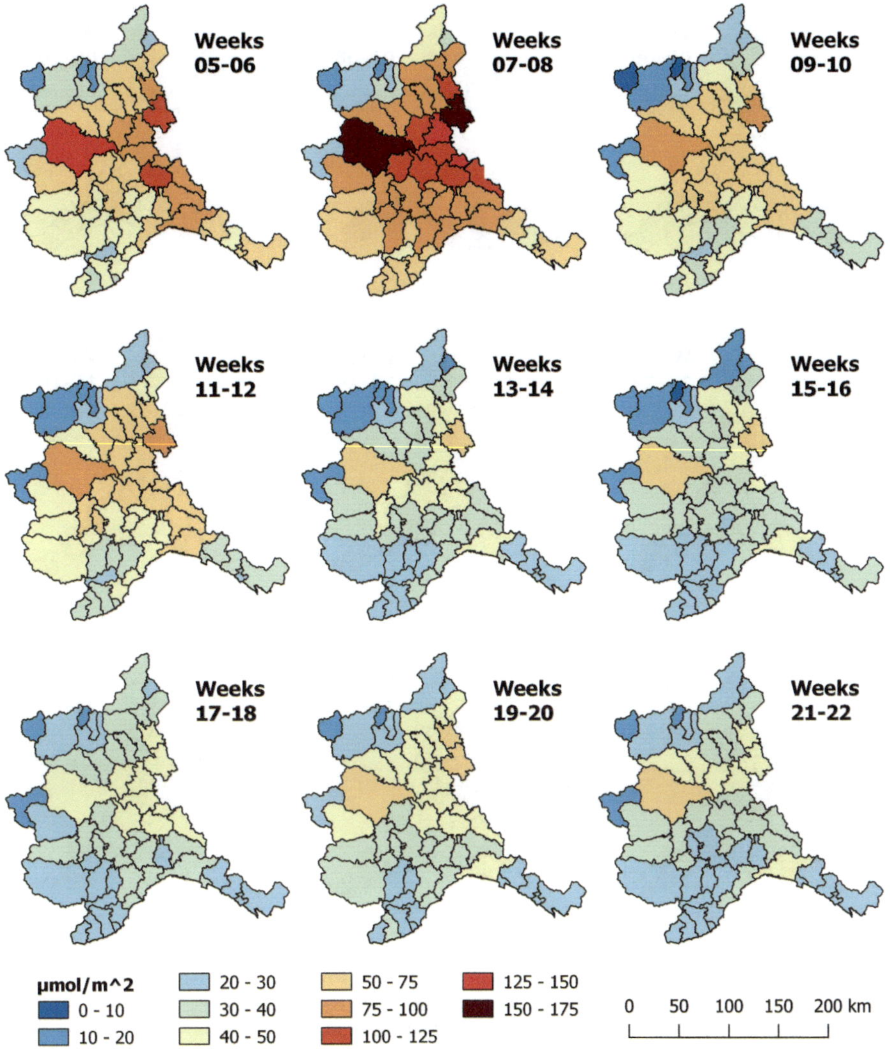

Fig. 7.4 Average concentrations of NO_2 for LLS

of high values far from it are the local system of Garessio and the axis Tortona–Turin. This pattern becomes more evident in the cartogram of NO_2 concentrations (Fig. 7.7b) where a group of local system with medium–high values on the eastern border is clearly visible. The values tend to decrease toward West except for Turin LLS which corresponds to the outlier value.

Not all variables present a clear pattern like the previous; however in the cartogram of the population density (Fig. 7.8a), it is possible to see a spatial correlation with the NO_2 concentrations, which is confirmed by a Pearson correlation index of 0.51 with

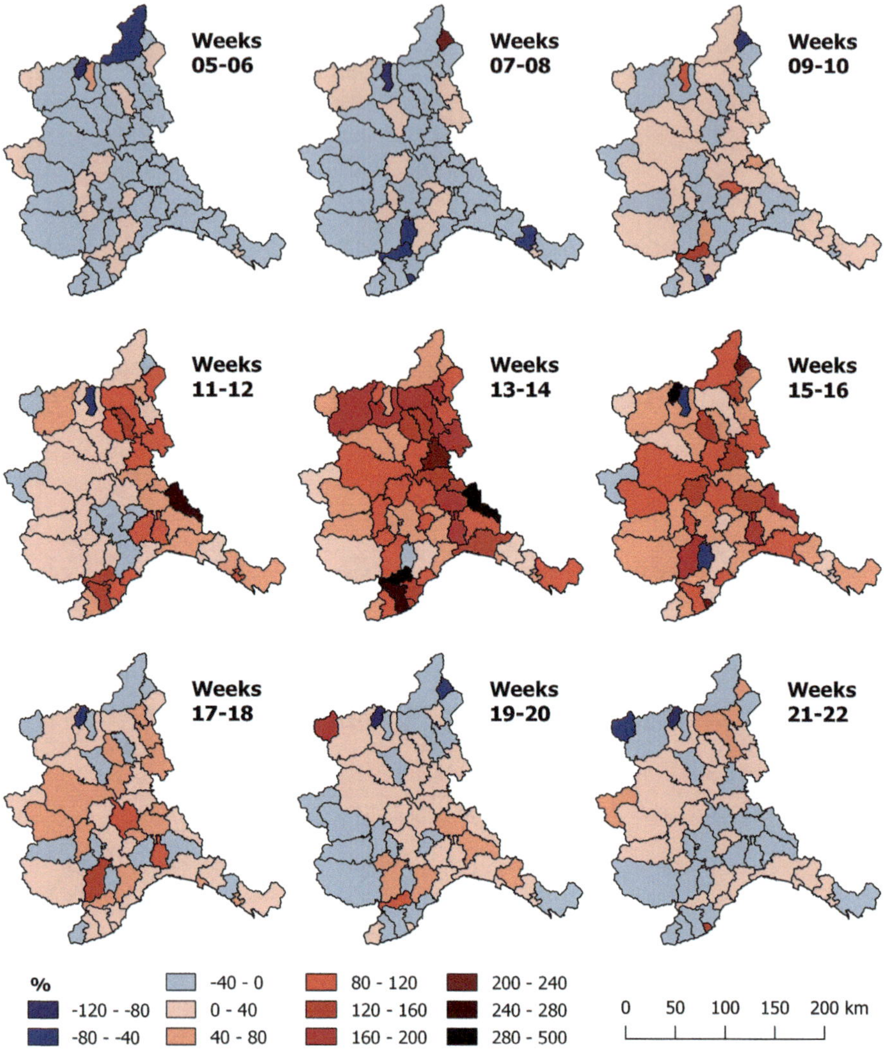

Fig. 7.5 Excess mortality for LLS

a *p*-value of 5.12e−05. Another pattern is shown in the cartogram of the proportion of the elderly population where the local systems of Liguria region are all part of the upper classes (Fig. 7.8b).

While, in the two cartograms of the self-containment of the job market indices (Fig. 7.8c, d) no scheme emerges but the local system of the regional capitals have the highest values in both indices, which is predictable as they are strong centers of attraction for labor.

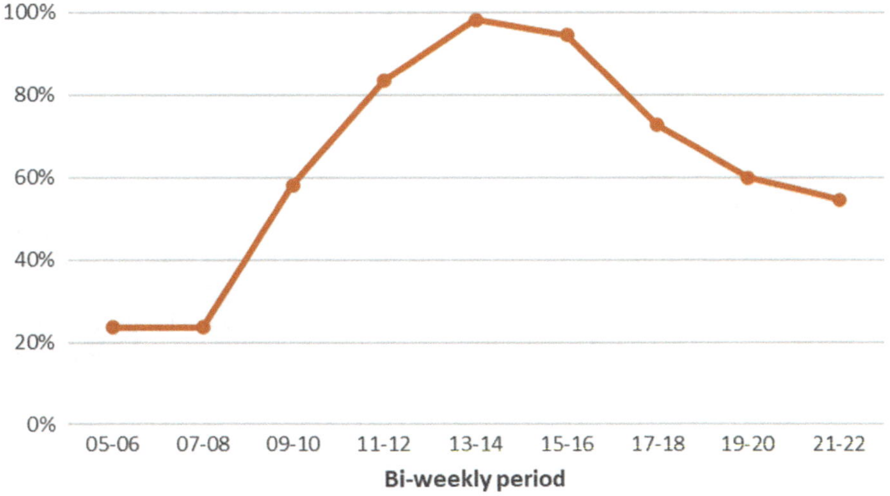

Fig. 7.6 Excess mortality trend

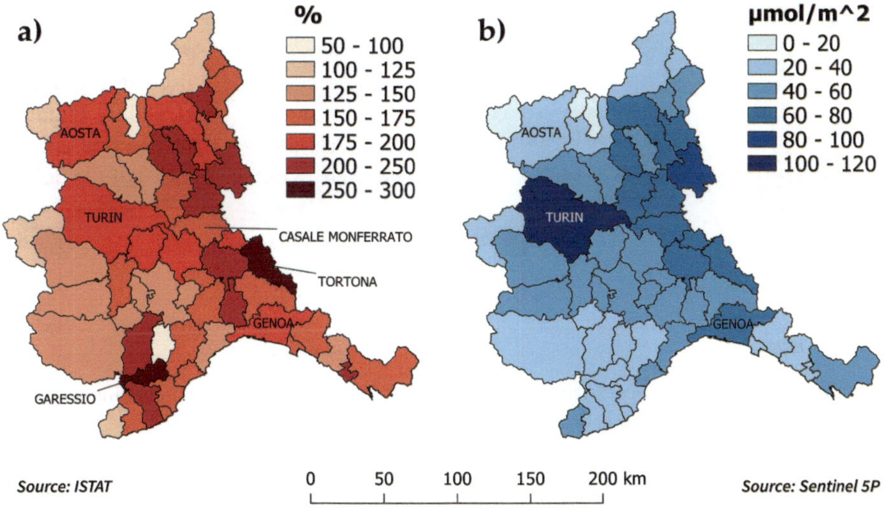

Fig. 7.7 **a** Excess mortality; **b** average concentrations of NO_2

7.3.2 Statistical Analysis

Spearman's correlation coefficient revealed that excess mortality during the first wave is moderately correlated to nitrogen dioxide concentrations ($\rho = 0.43$, *p*-value < 0.05), as evidenced by the regression line in the scatterplot in Fig. 7.9.

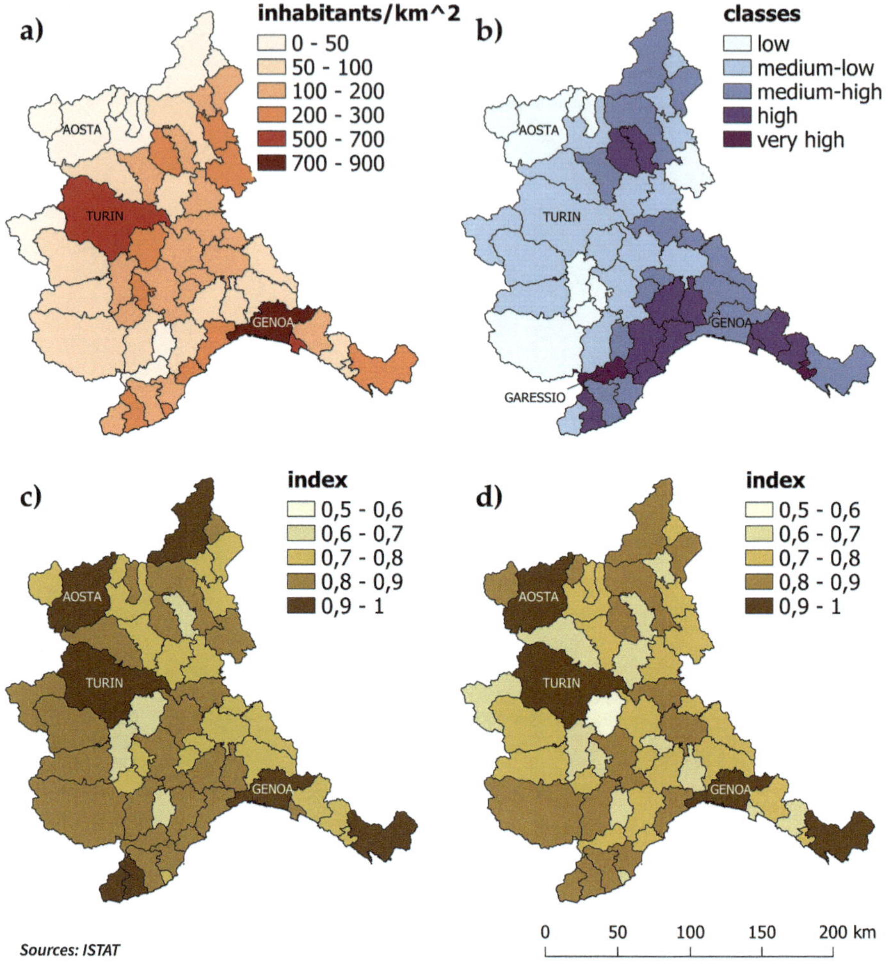

Fig. 7.8 **a** Population density (2020); **b** proportion of elderly population[3]; **c** self-containment index of the jobs demand (SCID); **d** self-containment index of the jobs supply (SCIS)

Subsequently, this correlation was also investigated using three quasi-Poisson regression models. Since the resulting coefficients of the regression models are reported on a logarithmic scale, they must be antilog to obtain rate ratios in the original scale for a better interpretation. Therefore, in the conversion process with the exponential function, some variables have been further transformed. The coefficients of NO_2 concentrations were multiplied for 10 so the rate ratio expresses the percent variation in excess mortality for each variation of 10 μmol/m^2. Similarly, other coefficients were transformed: the population density was multiplied for 100 so

[3] The classes were defined using the mean value and adding or subtracting one and two standard deviations.

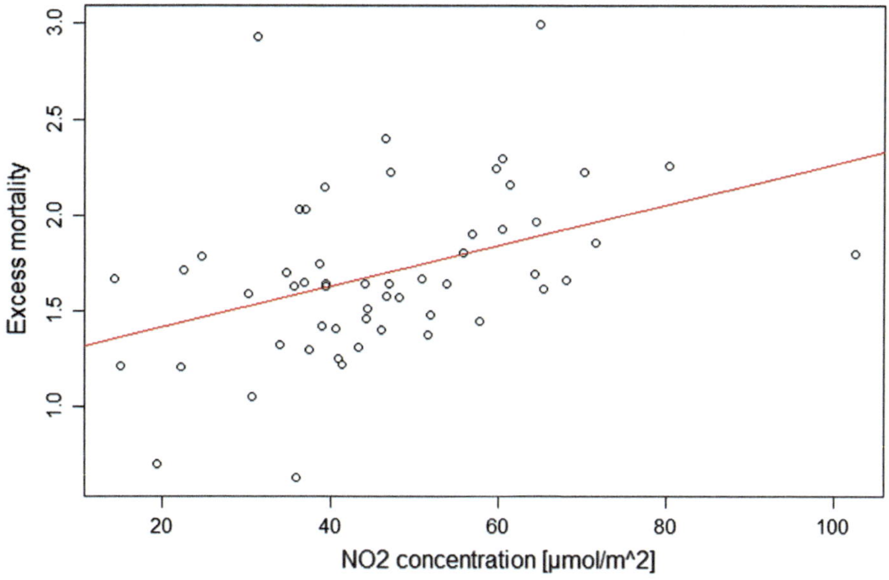

Fig. 7.9 Scatterplot: excess mortality versus NO_2 concentrations

Table 7.1 Values resulted from the regression in the first model (already retransformed)

Variables	Rate ratio	2.50%	97.50%	*p*-value
NO_2 concentration	1.01497	1.00332	1.02671	0.01450*

$***p < 0.001$; $**p < 0.01$; $*p < 0.05$

the rate ratio indicates the percent variation in excess mortality for a variation of 100 inhabitants/km^2. While, the coefficients of the two population proportions and the two self-containment indices (SCI) were divided by 100 so the rate ratios measures the percentage change in excess mortality for a one percentage point change in these indices.

In the first model, only the two main variables (NO_2 concentration and excess mortality) of the study were considered. As given in Table 7.1, each increase of 10 μmol/m^2 in NO_2 concentrations is associated with a 1.49%[4] (95%CI: 0.33 ÷ 2.67) increase in excess mortality.

Next to this, other variables were also considered for a possible confounding effect. These other two models differ for just one variable: one model considers the self-containment index of demand ("demand model") and the other the self-containment index of supply ("supply model"). In the "demand model" (Table 7.2), among all the variables, the only one statistically associated with the excess of mortality is the demand SCI variable.

[4] All the percentages are obtained subtracting 1 from the rate ratio values and multiplying it by 100.

Table 7.2 Values resulted from the regression in the "demand model" (already retransformed)

Variables	Rate ratio	2.50%	97.50%	*p*-value
NO_2 concentration	1.01813	0.99956	1.03714	0.06230
Population density	1.00314	0.98176	1.02489	0.77640
Over 65 population	1.02050	0.99696	1.04443	0.09380
Male population	0.98476	0.89891	1.07768	0.74150
Jobs per capita	2.46034	0.76704	7.94295	0.13770*
Demand SCI				

$***p < 0.001$; $**p < 0.01$; $*p < 0.05$

Table 7.3 Values resulted from the regression in the "supply model" (already retransformed)

Variables	Rate ratio	2.50%	97.50%	*p*-value
NO_2 concentration	1.01886	1.00081	1.03731	0.04643*
Population density	1.00352	0.98304	1.02434	0.73929
Over 65 population	1.02380	1.00064	1.04734	0.04886*
Male population	0.97373	0.88973	1.06459	0.56364
Jobs per capita	6.37255	1.61164	2.553444	0.01150*
Supply SCI	0.99352	0.98899	0.99806	0.00751**

$***p < 0.001$; $**p < 0.01$; $* p < 0.05$

Table 7.3 shows the rate ratios of the "supply model" in which four variables result associated with excess mortality. The associated percentage increase in excess mortality is 1.88% for each 10 μmol/m^2 increase in NO_2 concentration and 2.38% for one percentage point increase in the elderly population proportion. While, for an increase of one job per capita, the excess mortality is sixfold.

Both self-containment indices have an inverse association with the excess mortality indicated by the fact that both rate ratios are less than one. This inverse association means that each increase in one percentage point of the index is associated with a decrease in the excess mortality, respectively, of—0.65% for the demand SCI and—0.64% for the supply SCI.[5]

Moreover, in both two models, the variables of population density and the proportion of the male population do not find a significant association with excess mortality.

[5] In these cases subtracting one from the rate ratio resulted in the following calculations: "0.99341 − 1 = − 0.00659" for the SCID and "0.99352 − 1 = − 0.00648" for the SCIS.

7.4 Discussion and Conclusions

The relationship between NO_2 concentration and excess mortality investigated in this study partially emerges from the comparison of the respective cartograms. As previously described, these two variables tend to show higher values along the border with Lombardy region, particularly evident for the nitrogen dioxide pollutant. The zoning of NO_2 values follows the spatial distribution of the population density. This result is an expected consequence since more polluting sources (i.e., industries and vehicular traffic) tend to reside in the most populated areas.

The pattern of the highest excess mortality values near the Lombardy region needs more clarification: Lombardy region was the epicenter of the pandemic in Italy, and this explains how the local systems closest to the border have a higher excess of mortality, with a decreasing gradient toward West. However, this gradient is nonlinear as proximity is not the only contributing factor to the spread of the virus. In facts, an important element to consider is the degree of relations and connections that each local system has with other local systems. The more distant but more connected local systems, such as the Turin one, could be more affected by the spread of the virus than the systems closer to Lombardy but less connected, i.e., Casale Monferrato. For other local systems such as Garessio, the reason of high excess mortality value could be linked to the remarkably high value in the proportion of the elderly population. However, this kind of observations based on the comparison between only two variables could lead to misinterpretation of the real relationship; hence, statistical analyses are needed to consolidate preliminary observations. The nitrogen dioxide concentrations result associated to the excess mortality in two of the three regression models implemented. Our finding is therefore in accordance with previous studies (Filippini et al. 2021; Fattorini and Regoli 2020; Ogen 2020). In the "supply model", other variables, considered as confounders, showed a statistically significant association with the excess mortality. In the case of the proportion of the elderly population this result is congruent with the high incidence rate of COVID-19 in elderly people reported by the Italian National Institute of Statistic (ISTAT 2021b). Among the variables associated with excess mortality, the rate ratio with the greatest multiplicative effect is the one referred to the jobs per capita: since the period with higher excess mortality values coincides with the national lockdown, when the only movements allowed were mainly for work reasons. The number of jobs in a local system was critical for the spread of the virus because it increased the intensity of travels and the amount of close contacts between coworkers. Therefore, it can be assumed as the cause of such a great multiplicative effect. An unexpected result is the non-association of population density with excess mortality; in fact, a greater population density should correspond to a greater possibility of interpersonal contacts and therefore more infections. An interesting interpretation of this is provided by the study of Cremaschi et al. (2021) which suggests the adoption of a *relational density* (McFarlane 2016) to justify the spatial spread of the virus. Density can be considered a measure of how close social relations are in the resident population. The use of this variable would allow to highlight substantial differences in socio-spatial relationship that characterize small- or medium-sized territorial systems compared to those of large cities.

The two self-containment indices were used as a measure of the isolation of the local systems. Higher values correspond a smaller number of workers who arrive from outside the system or who leave the system in which they live to work outside. The inverse association with the excess mortality demonstrated by both indices means that a system with a higher SCI, net of other variables, recorded a lower excess mortality during the first wave of the pandemic. This relationship could be explained as lower exchanges of workers with other local systems. In an initial phase of the spread of the virus, when it was not yet homogeneously present on the territory, exchanges with external systems were essential for its transmission. Furthermore, considering that the only travels allowed during the lockdown period were those related to work, this fact strengthen the role of these indices in the interpretation of the spread of the disease.

The multivariate regression models carried out confirmed the existence of a significant statistical association between NO_2 concentrations and excess mortality values for local systems of the northwestern regions during the first COVID-19 epidemic wave in Italy.

The results obtained are thus part of a broader panorama of studies aimed at evaluating the effects of air pollution on human health (Carugno et al. 2016; Eum et al. 2019; Fattorini and Regoli 2020; Filippini et al. 2020; Ogen 2020) but also highlight the importance of conducting such analyses on a more appropriate geographic scale such as that of local systems.

Acknowledgements Here we want to thank Giovenale Moirano (MD-Ph.D., University of Turin) for the precious scientific contribution he has provided us.

References

Berk R, MacDonald JM (2008) Overdispersion and Poisson regression. J Quant Criminol 24:269–284. https://doi.org/10.1007/s10940-008-9048-4

Carugno M, Consonni D, Randi G, Catelan D et al (2016) Air pollution exposure, cause-specific deaths and hospitalizations in a highly polluted Italian region. Environ Res 147:415–424. https://doi.org/10.1016/j.envres.2016.03.003

CIESIN (2018). Gridded population of the world, Version 4 (GPWv4): population count, Revision 11. NASA Socioeconomic Data and Applications Center (SEDAC). https://sedac.ciesin.columbia.edu/data/set/gpw-v4-population-count-rev11

Ciminelli G, Garcia-Mandicó S (2020) COVID-19 in Italy: an analysis of death registry data. J Public Health 42(4):723–730. https://doi.org/10.1093/pubmed/fdaa165

Coccia M (2020) Factors determining the diffusion of COVID-19 and suggested strategy to prevent future accelerated viral infectivity similar to COVID. Sci Total Environ 729:138474. https://doi.org/10.1016/j.scitotenv.2020.138474

Conticini E, Frediani B, Caro D (2020) Can atmospheric pollution be considered a cofactor in extremely high level of SARS-CoV-2 lethality in Northern Italy? Environ Pollut 261:114465. https://doi.org/10.1016/j.envpol.2020.114465

Cremaschi M, Salone C, Besana A (2021) Densità urbana e Covid-19: la diffusione territoriale del virus nell'area di Bergamo. Archivio Di Studi Urbani e Regionali 131:5–31. https://doi.org/10.3280/ASUR2021-131001

DPCM (2020a) Decreto del Presidente del Consiglio dei Ministri 9 marzo 2020. Tratto da Gazzetta Ufficiale della Repubblica Italiana. https://www.gazzettaufficiale.it/eli/id/2020/03/09/20A01558/sg

DPCM (2020b) Decreto del Presidente del Consiglio dei Ministri 26 aprile 2020. Tratto da Gazzetta Ufficiale della Repubblica Italiana. https://www.gazzettaufficiale.it/eli/id/2020/04/27/20A02352/sg

Eum K-D, Kazemiparkouhi F, Wang B, Manjourides J, Pun V, Pavlu V, Suh H (2019) Long-term NO_2 exposures and cause-specific mortality in American older adults. Environ Int 124:10–15. https://doi.org/10.1016/j.envint.2018.12.060

Fattorini D, Regoli F (2020) Role of the chronic air pollution levels in the Covid-19 outbreak risk in Italy. Environ Pollut 264:114732. https://doi.org/10.1016/j.envpol.2020.114732

Filippini T, Rothman KJ, Vinceti M (2021) Associations between mortality from COVID-19 in two Italian regions and outdoor air pollution as assessed through tropospheric nitrogen dioxide. Sci Total Environ 760:143355. https://doi.org/10.1016/j.scitotenv.2020.143355

Filippini T, Rothman KJ, Goffi A, Ferrari F, Maffeis G, Orsini N, Vinceti M (2020). Satellite-detected tropospheric nitrogen dioxide and spread of SARS-CoV-2 infection in Northern Italy. Sci Total Environ 739. https://doi.org/10.1016/j.scitotenv.2020.140278

ISTAT (2015a) Sistemi locali del lavoro - Nota metodologica. Report statistiche. Roma, Istat. https://www.istat.it/it/files//2014/12/nota-metodolgica_SLL2011_rev2015a0205.pdf

ISTAT (2015b) Indicatori di qualità dei sistemi locali del lavoro. Roma. Istat. https://www4.istat.it/it/strumenti/territorio-e-cartografia/sistemi-locali-del-lavoro/indicatori-di-qualit%C3%A0-sll

ISTAT (2020a) Composizione dei Sistemi Locali del Lavoro 2011 a patire dal 1° gennaio 2020. Roma, Istat. https://www.istat.it/it/archivio/252261

ISTAT (2020b) I.stat, Popolazione residente al 1° gennaio. Roma. Istat. http://dati.istat.it/Index.aspx?QueryId=19101

ISTAT (2021a) Dataset con i decessi giornalieri. Roma, Istat. https://www.istat.it/it/archivio/240401

ISTAT (2021b) Impatto dell'epidemia covid-19 sulla mortalità totale della popolazione residente. V Rapporto, 5 marzo 2021. https://www.istat.it/it/files//2021/03/Report_ISS_Istat_2020_5_marzo.pdf

Martelletti L, Martelletti P (2020) Air pollution and the novel Covid-19 disease: a putative disease risk factor. SN Compr Clin Med 2(4):383–387. https://doi.org/10.1007/s42399-020-00274-4

McFarlane C (2016) The geographies of urban density: topology, politics and the city. Prog Hum Geogr 40(5):629–648. https://doi.org/10.1177/0309132515608694

Ogen Y (2020) Assessing nitrogen dioxide (NO_2) levels as a contributing factor to coronavirus (COVID-19) fatality. Sci Total Environ 726:138605. https://doi.org/10.1016/j.scitotenv.2020.138605

Pignocchino G, Pezzoli A, Besana A (2022) Satellite data and epidemic cartography: a study of the relationship between the concentration of NO_2 and the COVID-19 epidemic. In: Borgogno-Mondino E, Zamperlin P (eds) Italian conference on geomatics and geospatial technologies. Springer International Publishing, Cham, pp 55–67

Chapter 8
The COVID-19 Effects and the Development Process of Lanzo Valleys in a Metro-Mountain Perspective

Federica Corrado, Erwin Durbiano, and Gabriella Negrini

Abstract During the last century, the close relationship between the Lanzo valleys and the Turin conurbation has declined differently because of the specific local resources, networks, and the dominant development model, starting with the historic holiday resort of the Turin upper class up to the industrialization of the lower valley, linked to Turin industry and currently in crisis. Recent projects, carried out by the local community, however, testify to the desire for a new dynamism through innovative experiences that look beyond the traditional activities, as testified by the selection of this area in the first phase of the National Strategy for Internal Areas (SNAI). These dynamics have accelerated in relation to the COVID-19 pandemic which has further re-evaluated the role and importance of some local assets that have favoured the spread of a different model of living characterized by a temporary residency that positively exploits the qualities of an urban-mountain environment. On the other hand, there was a rediscovery of a different tourism, linked to places, history, and nature. This contribution deepens the evolution of these dynamics, focusing on the entire metropolitan-mountain area and on specific municipalities through qualitative-quantitative analyses to evaluate the effects both in the short and long-term allowing to establish which are, in the near future, the permanent assets on which to focus attention to trigger a leverage effect and which ones can be considered as temporary conditions that run out in a short time.

F. Corrado
Interuniversity Department of Regional and Urban Studies and Planning, Urban Technology and Planning, Politecnico di Torino, Turin, Italy
e-mail: federica.corrado@polito.it

E. Durbiano (✉)
Department of Regional and Urban Studies and Planning, Interuniversity Department of Regional and Urban Studies and Planning, Politecnico di Torino, Turin, Italy
e-mail: erwin.durbiano@polito.it

G. Negrini
CED PPN (European Documentation Centre on Nature Park Planning—DIST), Interuniversity Department of Regional and Urban Studies and Planning, Politecnico di Torino, Turin, Italy
e-mail: gabriella.negrini@polito.it

G. Brunetta et al. (eds.), *Post Un-Lock*, The Urban Book Series,
https://doi.org/10.1007/978-3-031-33894-6_8

Keywords Metro-mountain system · Mountain-urban relationships · Sustainable development · Protected areas

8.1 Introduction. Lanzo Valleys and the Metro-Mountain Perspective

In the past twenty years, the relationship between Lanzo Valleys and the urban area of Turin was expressed in multiple ways through a reinterpretation of local resources and their promotion, within local planning at different levels.[1] Nevertheless, the framework of design actions and the related recognition of strategic elements were put to the test both in the COVID-19 pandemic phase and in the current post-pandemic phase (German Environment Agency 2021).

The aim of this chapter is thus to investigate—through the analysis of institutional and territorial plans—the development processes and main strategic assets following the effects of the COVID-19 pandemic, considering the rise of a metro-mountain framework. To take on a metro-mountain perspective (Dematteis et al. 2017; Corrado 2021; Corrado and Durbiano 2018) means referring to new local configurations that abandon the classic city-countryside and city-mountain dichotomies to consolidate, instead, "urban" and "rural" as parts of the same system and elements that create territorial added value.

Lanzo Valleys area (Fig. 8.1) includes urban areas where the various services are concentrated (Lanzo, Ciriè) and that serve as the possible hubs of a metro-mountain system; an intersection essentially corresponding to the peri-urban Alpine area, in a strategic position with crucial access points of the Turin metropolitan area (Turin–Ceres railroad, Caselle Airport, SP 1 provincial road of the Lanzo Valleys from Turin to Balme, SP 2 from Turin to Germagnano) and, finally, a medium–high valley area that experienced, already in the pre-pandemic phase, a scattered rebirth related to the hybridization of tradition and innovation developed, in part, by new dwellers and, in part, by a new way of being mountaineers (Corrado 2014; Bender 2012).

---

[1] We hereby underline the importance of such a process in EU planning. On the topic, we mention the RURBANCE Alpine Space project (Final Report, 2015, *For a balanced development of the relations between rural and urban areas*), which puts a shared view of local development on the front stage by adopting a Joint Development Strategy able to overcome the limits of policy fragmentation and sectionalization, working on the integration of urban and rural aspects (Baschenis et al. 2015); the Alpine Space AlpBC project, which aims to capitalize on Alpine construction culture through smart regional planning and orientation strategies for sustainable development and circular economy in the Alps (Berta et al. 2015).

Fig. 8.1 Lanzo Valleys in the Turin metropolitan area

8.2 Research Methodology

To respond to the goals of this research, the analysis activities were articulated into four levels of investigation: the first level concerns the discussion of the reference framework of the broad-spectrum territorial institution plans currently in force; the second level concerns the strategic plan of the metropolitan area; the third level concerns the analysis of plans created in relation to national projects about inner areas (PNRR, national recovery and resilience plan 2021); the fourth level puts the attention on local mechanisms related to the pandemic and post-pandemic situation and its effects on the area.

The analysis phase was supported by a specific focus on the strategic role of protected areas for biodiversity, health, and well-being. Such protected areas, especially in the most marginal and fragile areas such as inner areas and mountains-are strategic assets to develop an integrated and sustainable territorial project, by promoting and establishing a network of natural, cultural, and identity-making heritage. Within this framework, the Lanzo Valley offers relevant, strategically protected areas connected to the metropolitan area that respond to the growing demand for sustainable and proximity tourism—well underlined by the COVID-19 pandemic—to develop metro-mountain enhancement policies that may provide a mutual benefit to the cities and valleys.

8.2.1 Elements Emerging from the Analysis

As for the first level of investigation, upon observation of the PTR (regional territory plan, Table 8.1) approved in 2011 and from the PPR (regional landscape plan, Tables 8.2 and 8.3) approved in 2017, we gathered what the large-scale paths for Lanzo Valleys area. Above all, the promotion of natural (mostly water, forest, and agro-silvopastoral) resources is in the scope of making Lanzo Valleys a regenerator of crucial ecosystem services for the entire urban/metropolitan network. Secondly, attention to the urbanization process, both in terms of latent potential and in terms of criticalities due to decommissioning, depopulation, and de-qualification of the built heritage. Finally, an implementation of actions to protect, safeguard, and promote the rich natural heritage of the area.

Table 8.1 Planning directions for Lanzo Valleys

Territorial components	Directions
Planning directions for Lanzo Valleys	
Environmental heritage	• Protection and management of water and forestry resources (special focus on medium–high mountain areas) • Fostering integrated production from renewable energy (hydroelectric and biomass)
Urban dynamics	• Redefining and acting upon the territorial rearrangement of the North quadrant of the Turin metropolitan area, with a role as external sub-pole • Limitation of widespread urbanization; reuse and recovery of abandoned and neglected areas
Specific rural elements	• Maintenance of a human presence and revalorization of mountain areas to live • Promotion of main natural resources (water, woods, pastures) even for production purposes
Tourism	• Favouring sustainable tourism • Promotion of historical/architectural and landscape heritage four tourism purposes
Transportation	• Redevelopment of the Turin–Ceres railroad • Facilitation of a fast connection with the Caselle Airport
Culture	• Promotion of natural (river), cultural, and landscape heritage • Connection of the historical/architectural heritage with the network of House of Savoy residences

Source PTR

Table 8.2 Ongoing dynamics of Lanzo Valleys

Territorial components	Ongoing dynamics
(a) Ongoing dynamics of Lanzo Valleys	
Environmental heritage	• Retreat of Glaciers • Abandonment of anthropogenic forest areas (chestnut trees) • Abandonment of marginal pastoral areas and consequent ecological restoration of the trees and shrubs
Urban dynamics	• Reckless and extensive construction • Inappropriate construction in terms of size and architectural language • Initiatives for the recovery of public spaces and small villages
Tourism	• Reduction of stable mountain tourism, also because of poor accessibility • Growth of touring and trekking activity • Promotion of the historical trails of mining and metalworking activity
Specific rural elements	• Tendency to abandon marginal pastoral areas • Mountain abandonment and territorial maintenance practices

Table 8.3 Strategic paths for Lanzo Valleys

Territorial components	Strategic paths
(b) Strategic paths for Lanzo Valleys	
Environmental heritage	• Active and multipurpose management of forest heritage • Promotion of forms of technical support to Alpine culture
Urban dynamics	• Renovation of Viù territory, considering the chaotic urbanization of the '70s and '80s
Transportation	• Promotion of the historical railroad
Specific rural resources	• Preservation of the historical/built and recycle heritage and the related rural contexts
Tourism	• Definition of network with initiatives of Turin metropolitan area (e.g. House of Savoy residence network) • Promotion of the cultural aspects of the fundamental and identity-making activities of the area and its production activity

Source PPR

Upon analysing the Strategic Plan of the Metropolitan City of Turin in the second level of investigation, the relationship between the area and the Turin conurbation or—even more—with the entire metro-mountain system emerge. Such a relationship is essentially expressed in three directions:

- *Digitalization, innovation, competitiveness, and culture*: increase the ability to create value in the different economic sectors (agriculture, tourism, services through cooperation in business networks, creation of supply chains, promotion of the area and its products);

- *Infrastructures for sustainable mobility*: take advantage of the current transformation of commuting cycles to improve the connectivity and accessibility to and of the metropolitan area;
- *Inclusion and cohesion*: promote equal opportunity for personal and community development in the different parts of the metropolitan area.

The third level of investigation allowed us, instead, to have an outlook of local projects as outlined in the PSL (local development plan), approved in 2014. The GAL (local action group) of Lanzo Valleys, Ceronda, and Casternone, which has drafted the plan, identified three lines: sustainable and accessible tourism; promotion of local architecture and landscape; development and innovation of local supply chains and production systems. These lines have been additionally implemented through the improvement of local governance in order to overcome the-present territorial fragmentation, through the enhancement of the tourist offer—as the pandemic has underlined—and through the definition of a sustainable forest sector.

Many of such elements in the projects are included in the SNAI (national strategy for inner areas) designed in 2013 and developed by the *Agenzia per la coesione territoriale* (agency for territorial cohesion). The SNAI has selected 19 municipalities belonging to the Unione Montana di Comuni delle Valli di Lanzo, Ceronda e Casternone and Unione Montana Alpi Graie (mountain unions of the Lanzo Valleys-Ceronda-Casternone and Graian Alps) as the beneficiaries of government resources through specific framework agreements. The aim is to reduce marginality of this area through multiple and integrated actions related, in general, to the improvement of essential services and the redefinition of the relationship with the Turin metro-mountain area. The main expected results of the strategy include the inversion of the negative demographic trend and the implementation of public services (health care, education, mobility, and digital connectivity). In order to reach this aim, certain actions have been taken as summarized in Table 8.4.

In the aims of PNRR, the municipalities of Lanzo Valleys have growingly focused on the "green revolution and environmental transition" trajectory, acting—in particular—on the infrastructural and technological adjustment of the Canavesana and Turin–Ceres lines. A renewed metropolitan railroad system, with a more relevant role for the valley terminal stations, may reduce travel time even from the Caselle airport.

Finally, the fourth level of analysis highlighted the territorial dynamics of the pandemic and post-pandemic periods. Throughout the summer of 2020—the year of the pandemic crisis—the outdoor sector witnessed a drop in tourism activities, a reduction in foreign tourism, and support to demand mainly related to proximity tourism. During the same period, it emerged from the OTR (Regional Tourism Observatory, 2021) data Piedmont Region recorded + 19% in arrivals and + 23% in overnight stays in the mountain area in the June–July 2020 period. Also in the past few years, the number of accommodation facilities in Lanzo Valleys was essentially stable in municipalities, with a maintenance of the existing structures or an increase in certain locations: 7 accommodation facilities and 31 beds were added in Ala di Stura over the 2018–2020 triennium; Groscavallo added 2 new facilities and 16 beds.

Table 8.4 Strategic action for Lanzo Valleys (SNAI)

Cornerstone	Context	Strategic action	
Essential services	Health care	1	Activation of community healthcare services
		2	Enhancement of the emergency management system
	Education	3	Consolidation of the link between the school and territory
		4	Networking of the local education system
	Mobility	5	Reorganization of the local public transportation system and connection with other locations
Local development		6	Recovery and promotion of the local supply chain
		7	Promotion of tourist, cultural, and environmental resources
		8	Development of digital services
Technical support and co-design		9	Technical support and co-design

The pandemic has thus underlined the fragility and the need for such areas both in terms of renovation of the tourist sector and in terms of design, with a certain degree of regard towards the most vulnerable groups.

8.2.2 Focus on the Area: The Protected Areas for Biodiversity, Health, and Sustainable Development

Protected areas are an essential resources for biodiversity, ecosystems services conservation, health, and well-being, and for a sustainable territorial project (SDGs, UN Agenda 2030 2015, CBD 2010–2020, CBD Post-2020 Biodiversity Targets, EU Biodiversity Strategy for 2030), and also given the challenges presented by the great environmental and social changes (Dudley et al. 2017) and the international sustainability goals (UN SDGs).

As special areas that are "custodians" of the precious natural and cultural territorial value system (Gambino and Peano 2015), protected areas play a strategic role in the development of cultural/social/economic sustainability and in the enhancement policies of the park territory and its context, especially in the most fragile, inner, and marginal areas, to respond to a growing demand for nature and sustainable and proximity tourism, as suitably underlined by the post-COVID-19 emergency (Voghera et al. 2019, 2021; Hockings et al. 2020).

Especially in urbanized areas, nature parks and protected areas also confirm themselves as well-established "laboratories" for experimentation of integrated conservation and development policies and careful projects exportable to the wider territory for local communities, in line with the "new conservation paradigms" affirmed in Durban in 2003 (IUCN 2003 "Benefit beyond boundaries", Phillips 2003).

From an integrated conservation and development perspective, sustainable tourism—widely affirmed in protected areas at international level in its various forms of ecotourism and nature-based tourism (Leung et al. IUCN 2018; ECST-European Charter for Sustainable Tourism, EUROPARC Federation 1991 and 2010, IUCN WCC Marseille Resolution 130-Strengthening sustainable tourism's role in biodiversity conservation and community resilience)—is indeed among the most effective sectors for nature parks and protected areas to play the role as strategic nodes of a wider territory and potential drivers for development (Peano 2013): key assets to invest in to trigger promotion and enhancement processes and dynamics and create networks of resources, subjects, and projects, contributing to reverse the trend of weakening, marginalization, and abandonment that often characterize mountain, inland, and fragile areas by favouring, instead, the taking-care of territories, in line with the SNAI and other initiatives and projects developed in this direction at national level (*APE. Appennino Parco d'Europa* project, the "ParchiAperti" Platform -Federparchi and other partners initiative; "Sentiero dei Parchi"-MATTM, CAI-Club Alpino Italiano; *Manifesto of Camaldoli* promoted by SdT-Società dei Territorialisti; MATTM, ASTER 2013 Report and MATTM 2015 Report).

In particular, the European Charter for Sustainable Tourism (ECST), launched by EUROPARC Federation in 1991 and subscribed by numerous European protected areas, offers great relaunch opportunities for the territories including protected areas as a methodological and certification tool—both for nature parks/protected areas and local tourism operators and businesses—based on a voluntary agreement between all the local actors and communities interested in creating and implementing a common strategy.

In this framework, the study area shows significant potential and opportunities in the perspective of a desirable alliance between mountain and city that may benefit both. An important natural, landscape, historical/cultural, and identity-making heritage with different features in the three Valleys (Viù Valley, d'Ala Valley, Val Grande Valley—the latter subject to the proposed creation of a natural regional park "Alta Val Grande") as well as a great variety of contexts, from the wilder pre-Alpine areas to the rural, urban, and peri-urban contexts. The territory also boasts a solid tradition and vocation for natural, cultural, and proximity tourism (a Turin tradition of mountain climbing ever since the late-1800s) that satisfies the growing demand for nature, health, and quality tourism close to the city. It is a high-quality urban-mountain environment that boasts a set of protected areas including Nature Parks, Nature Reserve, Natura 2000 Network Sites, part of the Royal Parks system including La Mandria Nature Park (a "green lung" of the metropolitan city of Turin). These resources are key elements in the "Corona Verde" (green crown) strategic project, subject to intense planning, management, and design activity coordinated by the regional administration. Considering the Lanzo Valleys, are of particular interest the Ponte del Diavolo Nature Reserve—a place of special historic, geological, and symbolic value—and the Zona di Salvaguardia della Stura di Lanzo River (part of the Stura di Lanzo SCI—Site of Community Interest), whose territorial contiguity allows the presence of an important ecological corridor. In addition, there are the SCIs of Grotte di Pugnetto and Pian della Mussa, the Colle del Lys provincial Nature Park (especially interesting from a historic/cultural, architectural, and landscape values), the Vauda Nature Reserve, and the Monte Lera Nature Reserve.

A set of strategic resources for the enhancement of the territory as well as sustainable and outdoor tourism, which may be noticed from the existence of trails and projects, in some cases promoted in partnership with the protected areas (e.g. the **Ve.La** project, which the Ente Parchi Reali is a partner of). In particular, we hereby report the creation of an articulate network of cycling/walking paths, tourist itineraries, and trails connecting the Lanzo Valleys with the metropolitan context, including the **Corona Verde Stura**, a network of cycle/pedestrian paths linked with the "Corona di Delizie" Royal Residence network and Ciclostrade della Stura di Lanzo cycling/walking itineraries; the **StouRing** project, a circular route of mixed-use (cycling, walking, touring) paths joining Lanzo, Venaria, and Borgaro, as well as involving the Vauda and the Ceronda and Casternone valleys; the **Ve.La** project—in progress—that is a 125 km (78 mi) itinerary of cycle paths along the Sturadi Lanzo River in the Lanzo Valleys and the "Corona di Delizie", in continuity with Turin and the River Po cycle path; the **GTA (Great Alpine Crossing)** in the Lanzo Valleys; the **Tour della Bessanese** in the Northern Graian Alps; the **Anello Ceronda MTB (ACM)**, a cycling/touring itinerary starting from Lanzo Torinese, crossing the high points of the Stura di Lanzo River, climbing back up the valleys of the Ceronda and Casternone rivers, then reaching the trails around Colle del Lys.

The features and values referenced offer a significant opportunity for the development and enhancement of the territory in a metro-mountain view through the activation of integrated strategies and system policies, the promotion of innovative forms of sustainable tourism related to places, history, nature, and the development of innovative housing models in the perspective of a restored balance of the relationships between the mountain and city.

8.3 New Metro-Mountain Relationships for Local Sustainability

Metro-mountain projects developed in this area and beyond (consider in Susa Valley) represent an increasing effort to redefine development policies and programmes within a local framework that appears growingly complex and multi-level (Corrado 2021). The elements identified and investigated in this research study show how the COVID-19 pandemic provided a new lever for relevant change in the local production and living sectors. Such fields growingly intersect and give life to new forms of urbanity that find expression in the experimentation of innovative activities and forms of residency (Perlik 2011) focused on quality of life and natural heritage (Keller 2010).

In relation to the pandemic period, the various local stakeholders have grown the consciousness of the importance to network on the patrimonalization process, or as Bonomi (2020) said, they have grown more eager to desire of local community to which the subjects belong. Therefore, adopting a proactive approach strengthens the strategic value of local dynamics and becomes the core of "territorial care" (Magnaghi 2021). The final goal is to reach a "transformative resilience": one based on overcoming the temptation to return to "how it was before" or a restart based merely on quantitative growth (Cortile dei Gentili Scientific Committee 2020), by implementing actions to:

- intensify and redefine the metro-mountain relationship by working on proximity and interdependence, with a special focus on the experimentation of forms of governance and local agreements;
- improve accessibility to reduce the marginality of inner areas through the strengthening and greater efficiency of local transportation (modernization of the Turin–Ceres railroad and implementation of terminal stations in the valleys, innovation in car-sharing and carpooling systems);
- develop more attractive living policies through actions to highlight the values and quality of the territory (environment, socialization, landscape) and improve the related policies (essential services, energy-forestry policies, real estate requalification);
- work on an idea of a local laboratory to support innovation in various sectors—from soft tourism to green economy, etc.;
- capitalize the co-design results with the community to respond efficiently and effectively to the internal/external demands of the system.

References

Baschenis G, La Greca F, Quarta M (eds) (2015) Rurbance. Rural urban governance. Piedmont Regional Office, Turin

Bender O (2012) New immigration into the European Alps: emerging research issues. Mt Res Dev 32(2)

Berta M, Corrado F, De Rossi A, Dini R (2015) Architettura e territorio alpino. Scenari di sviluppo e di riqualificazione energetico-edilizia del patrimonio costruito. Piedmont Regional Office, Turin

Bonomi A (2020) Il territorio come costruzione sociale al tempo del Covid. Scienze del Territorio, Special issue, PUF, Florence

Corrado F (ed) (2021) Urbano Montano, verso nuove configurazioni e progetti di territorio. Franco Angeli, Milan, pp 1–20

Corrado F (2014) Processes of re-settlement in mountain areas. J Alpine Res 102(3)

Corrado F, Durbiano E (2018) Alla ricerca di un (ri)equilibrio tra montagna e città, Politiche Piemonte, ISSN 279-5030

Dematteis G, Corrado F, Di Gioia A, Durbiano E (2017) L'interscambio montagna città- Il caso della città metropolitana di Torino. Franco Angeli, Milan

Dudley N, Ali N, MacKinnon K (2017) Natural solutions. Protected areas helping to meet the sustainable development goals. IUCN, WCPA

Europarc Federation (1991, revised 2007, 2010) European Charter for sustainable tourism in protected areas. EUROPARC Federation, Regensburg, Germany

Gambino R, Peano A (eds) (2015) Nature policies and landscape policies. Towards an alliance. Springer, Dordrecht

German Environment Agency (2021) The impact of COVID-19 on economies in the Alpine region. Umweltbundesamt, Dessau-Roßlau. http://www.umweltbundesamt.de/publikationen

Hockings M, Dudley N, Elliott W, Napolitano Ferreira M, MacKinnon K, Pasha MKS, Phillips A et al (2020). Editorial essay: COVID-19 and protected and conserved areas. In: Hockings, M. PARKS, 26(1). IUCN WCPA, Gland, Switzerland

Keller L (2010) Quality of life in the Alps. Developing a scientific model on an elusive geographic topic. In: Borsdorf A et al (eds) Challenges for mountain regions—tackling complexity. Bohlau, Wien

Leung YF, Spenceley A, Hvenegaard G, Buckley R (eds) (2018) Tourism and visitor management in protected areas. Guidelines for sustainability. Best Practice Protected Area Guidelines, Series No. 27, Gland, Switzerland

Magnaghi A (2021) Il principio territoriale. Bollati Boringhieri, Turin

Ministero dell'Ambiente e della Tutela del Territorio e del Mare (MATTM), Fondazione per lo sviluppo Sostenibile (2015) La Carta di Roma e i Parchi Nazionali, Primo rapporto sulle sinergie tra Capitale Naturale e Capitale Culturale, Roma

Ministero dell'Ambiente e della Tutela del Territorio e del Mare (MATTM), Fondazione per lo sviluppo Sostenibile, Federparchi, UNIONCAMERE, Consorzio AASTER (2013) Parchi come luogo di incontro tra green economy e green society, Research reports

Peano A (2013) I nuovi paradigmi della conservazione della natura. In Convegno ParchiBellaImpresa, Convegno sul sistema di gestione delle Aree Protette Torino, 28 March, Museo Regionale di Scienze Naturali, Turin

Perlik M (2011) Alpine gentrification: the mountain village as a metropolitan neighbourhood. J Alp Res 99(1)

Phillips A (2003) Turning ideas on their head the new paradigm for protected areas. J George Wright Soc 20(2), The George Wright Forum, Hancock, Michigan

Voghera A, Negrini G, Sammuri G, Agostinelli A (2021) Il ruolo delle Aree Protette per la ripartenza del Paese, in Urbanistica Informazioni, 265. INU Edizioni, Rome

Voghera A, Negrini G, Salizzoni E (2019) Parchi e Paesaggio. Dalla pianificazione all'efficacia di gestione, in Rapporto dal Territorio 2019, INU Edizioni, Rome

Chapter 9
Analysis of Hydrogeological Risks Related to Climate Change: Testing the *ClimeApp* Assessment Tool on the Torino Nord Homogenous Zone

Elena Pede, Mattia Scalas, and Luca Staricco

Abstract The current changes in temperatures and precipitations can lead to increased frequency and magnitude of natural hazards such as floods, resulting in forthcoming losses of life and damages to private and public properties. This paper presents an applied hydrogeological risk assessment methodology developed as part of an interdisciplinary European project between France and Italy (Interreg Alcotra ARTACLIM). A practical framework is proposed to assess the risk for urban settlements and infrastructural assets at a sub-regional level, within one of the 11 homogeneous zones of the Turin Metropolitan City, the "Torino Nord" Homogeneous Area of Pinerolo (ZOP). Based on the most reliable guidelines of institutions and organizations such as IPCC, United Nations, Covenant of Mayors for Climate & Energy EUROPE, the methodology here presented identifies risks related to climate change as a function of hazard (H), exposure (E) and vulnerability (V) factors, the latter one being in turn a function of sensitivity (S) and adaptive capacity (AC). Each factor is operationalized through specific indicators. The results of the analysis generate a geo-localized risk score that can be used to support urban planners and local policymakers to prioritize the adaptation measures required for reducing hydrogeological damages related to climate change.

Keywords Risk assessment · Climate change · Hydrogeological risk · ClimeApp · Turin Metropolitan Area

---

E. Pede (✉) · M. Scalas · L. Staricco
DIST—Interuniversity Department of Regional and Urban Studies and Planning, R3C—Interdepartmental Responsible Risk Resilience Centre, Politecnico di Torino, Turin, Italy
e-mail: elena.pede@polito.it

G. Brunetta et al. (eds.), *Post Un-Lock*, The Urban Book Series,
https://doi.org/10.1007/978-3-031-33894-6_9

9.1 Introduction

It is well-acknowledged that climate change can lead to increased frequency and magnitude of natural hazards such as landslides, rock falls, debris flows, avalanches, forest fires and floods (Field et al. 2012), which can generate potential negative impacts on infrastructures, socio-economic and cultural activities, ecosystems and so on. Most of these impacts—floods, in particular—cannot be effectively faced at the local level, but require conceiving and implementing adaptation strategies at a sub-regional, inter-municipal level (Birkmann 2011).

This article presents *ClimeApp*, a spatial risk assessment tool developed in the framework of the European Interreg research project ARTACLIM (http://artaclim.eu) to assess the main challenges related to natural hazards due to climate change at the sub-regional level. *ClimeApp* allows to evaluate how climate change can increase risks in terms of natural hazards (floods, landslides, avalanches, forest fires and droughts) on built settlements, infrastructures, tourism, agriculture, forests and biodiversity. *ClimeApp* assessments can be developed with reference to two representative concentration pathway (RCP) scenarios (the IPCC RCP4.5 and RCP8.5 scenarios; see IPCC 2014) and two time periods (2021–2050 and 2070–2100).

In the next sections, after a short presentation of the methodology at the basis of the tool, a pair of *ClimeApp* experimentations will be illustrated. They concern two of the eleven so-called *homogeneous zones* in which the Metropolitan City of Turin (chief town of the Piedmont region) is articulated; this spatial scale was chosen as it is often considered the more suitable administrative dimension to elaborate ad hoc adaptation strategies and measures. The two experimentations will show how *ClimeApp* can be used as a spatial decision support system for public administrations and local stakeholders in order to identify how hydrogeological risks due to climate change are (and will be in future) distributed inside a homogeneous zone, and which adaptation measures can be implemented to face these expected risks for built settlements and infrastructures.

9.2 The ClimeApp Assessment Tool

ClimeApp is based on the guidelines developed by institutions and organizations such as IPCC, United Nations, Covenant of Mayors for Climate & Energy EUROPE and the Alpine Convention (Bertoldi 2018; Probst et al. 2013; Schindelegger and Kanonier 2019; Zollner 2018), which identify risks related to climate change as a function of *Hazard* (H), *Exposure* (E) and *Vulnerability* (V) factors, the latter one being in turn a function of *Sensitivity* (S) and *Adaptive Capacity* (AC).

For the risks, we are considering in this paper, e.g., hydrogeological risks for built settlements and infrastructures, the *Hazard* represents the past and future occurrence of hydrogeological events due to the changes in the temperature and rainfall patterns (Oppenheimer et al. 2014). The *Exposure* refers to the built settlements and to the infrastructural systems that can be hit by these hydrogeological events. *Sensitivity* measures the degree to which the built and infrastructural systems can be unfavourably or beneficially affected by a certain climate change (GIZ 2017), while *Adaptive Capacity* refers to the ability of those systems to adapt to this change and its consequences (Oppenheimer et al. 2014).

Each factor is operationalized through specific indicators,[1] which are calculated for each of the municipalities within the considered homogeneous zone. In order to be aggregated in *Hazard*, *Exposure* and *Vulnerability* indexes, all indicators are normalized by applying the Min–Max method (Ellena et al. 2020; GIZ 2017), which translates their values into scores between 0 and 1. For some indicators, lower values reflect positive conditions; in this case, an operation of *conversion* is made by reversing the direction of the value range in order to have all indicators in the same normalization ranking (from 0 to 1). *Hazard*, *Exposure*, *Sensitivity* and *Adaptation Capacity* indexes are calculated as a weighted average of their normalized indicators; *Vulnerability* is calculated as the average of *Sensitivity* and *Adaptation Capacity*. At the end, the final risk value for each municipality is calculated as the product of the values of the *Hazard*, *Exposure* and *Vulnerability* indexes.

A *Geographic Information System* (GIS) spatial analysis tool is provided by *ClimeApp* to map and represent the distribution of the values of the indicators and indexes inside the considered homogeneous zone.

9.3 Testing *ClimeApp* on the Homogeneous Zone of Pinerolo

The *ClimeApp* tool has been tested as first instance in the framework of the ARTACLIM project on the *Homogeneous Zone of Pinerolo* (HZP), due to its geographical structure and the economic vocation based on outdoor sports (Ellena et al. 2020). From a geomorphological point of view, the HZP can be divided into three parts: an agricultural plain (12 municipalities), a hilly belt around the main town of Pinerolo (10 municipalities) and a mountainous area which borders on France and includes three valleys lying end to end (Val Pellice, Val Germanasca and Val Chisone) (23 municipalities).

Observing the climatic variations, results highlight a general increase in temperatures, more pronounced in the long-term period with the RCP 8.5 scenario. Looking

[1] For a detailed description of the indicators used in the case of hydrogeological risks for built settlements and infrastructures, see Ellena et al. (2020). Hazard indicators measure, in particular, the increase of temperature and precipitation frequency (through, for example, the annual total precipitation in wet days, the total number of days when precipitation > 1 mm, etc.). Examples of exposure indicators are the number of residents (or the hectares of industrial and commerce buildings) in places that could be negatively affected by floods. Sensitivity can be assessed using indicators such as the percentage of buildings in bad condition on the total number of buildings at the municipal level, or the number of inhabitants over 75 years old on the total population. The number of voluntary organizations active at municipal level and the percentage of dwelling units served by ultra-wide band technology are example of indicators for the adaptive capacity.

back to the phase 1981–2010, the observed climate evolution in terms of precipitation regimes and thermal increase has affected mainly the municipalities in the lower valleys, while it has had little impacts in the mountainous area until now. In future, projections may follow opposite trajectories. A strong reduction in the number of days with minimum and maximum temperature lower than 0 °C is expected over the area, but more pronounced in the high-elevation municipalities. Besides, the climate models predict a reduction in the number of rainfall days, more pronounced in the Alps, but an increase in precipitation intensity especially over the plains.

The final result is an overall increase of hydrogeological instability over the HZP with different levels of risk according to the kind of natural hazard addressed, the temporal frame considered and the three indexes' values (*Hazard*, *Exposure* and *Vulnerability*, Fig. 9.1). Therefore, these risk levels distribute very differently depending on elevation and morphology.

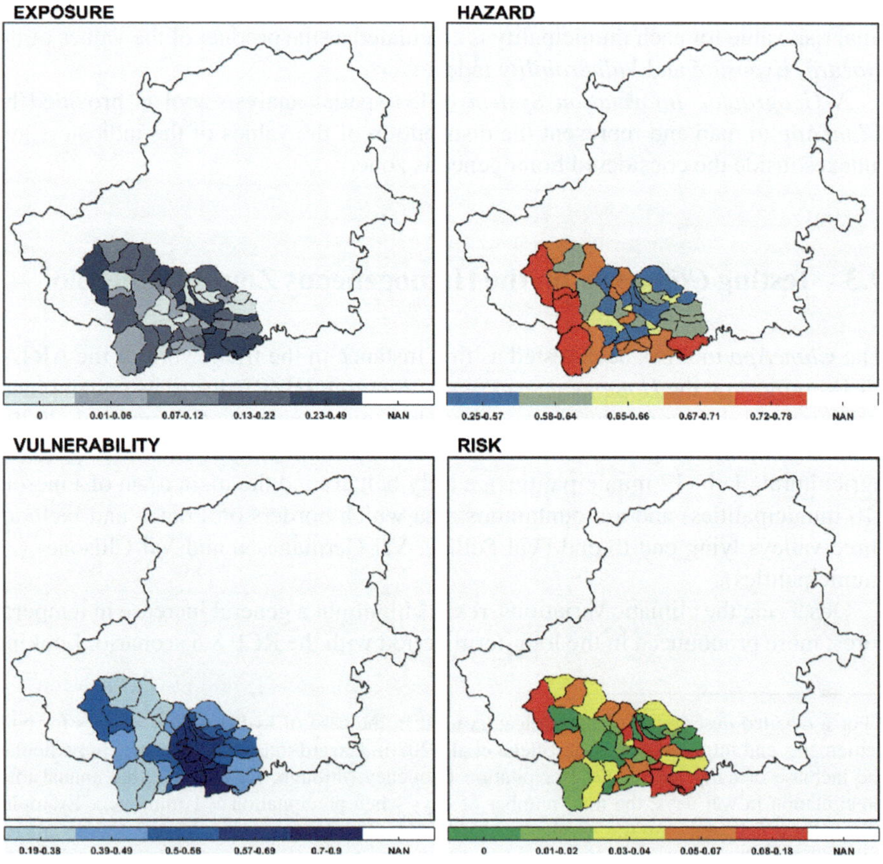

Fig. 9.1 From top left: exposure, vulnerability, hazard and risk (RPC4.5, 2021–2050, settlement and flooding) in HZP from ClimeApp visualizations

With reference to floods, settlements in high mountains and hilly territories will be increasingly vulnerable because of poor building conditions, high percentages of land use consumption and low adaptive capacity levels.

Flat and low hill areas of the HZP, however, will be most affected (specifically, municipalities located between the confluence of the Chisone and Pellice rivers). Even if a modest increase is expected in rainfall frequency and intensity (hazard), the exposure will be high due to the number of buildings and infrastructure systems considered at risk of flooding and to the number of inhabitants located in the same areas.

Finally, in the mountain area the hydrogeological aspects are complicated by the instability of the slopes and the increase of avalanches.

The significant rise of temperatures and the changes in the rainfall regimes will also have indirect consequences. The local economy is mainly based on outdoor sports and active tourism; being mainly dependent on natural resources and climate, this key sector is therefore particularly vulnerable to impacts of climate change.

In particular, a reduction in the frequency and intensity of snowfall will force the tourism sector to concentrate skiing activities at altitude over 1500 m. The negative impacts could be partially balanced for the high-elevation municipalities by increased opportunities related to summer tourism. The limited expected temperature increases and the decreases in precipitation, indeed, will extend the typical length of the summer season and, at the same time, the hills and the plain will be more hit by heat waves leading more people to look for coolness at a higher altitude.

However, these effects could be exacerbated by the increase of hydrogeological instability. This means that even if the touristic municipalities in the mountain part of the HZP are not significantly affected by climatic risks, they can be indirectly harmed in terms of accessibility of the infrastructural system due to the intensification of extreme events.

9.4 The Application of *ClimeApp* to the Homogeneous Zone of North Turin

9.4.1 Context and Methodology

In the framework of the Post Un-Lock project, a second application of *ClimeApp* was carried out on the Homogeneous Zone 4 of North Turin (HZNT). It includes the seven municipalities of Borgaro Torinese, Caselle Torinese, Leinì, San Benigno Canavese, San Mauro Torinese, Settimo Torinese and Volpiano. Since 2017, these municipalities have been joined by a new one, Mappano, which was instituted as a result of the separation of portions of the territory of Caselle Torinese, Borgaro Torinese, Settimo Torinese and Leinì.

HZNT is one of the smallest homogeneous zones in the Metropolitan City of Turin, both in terms of surface area (165.3 km^2) and number of component municipalities.

However, it is densely populated (129,721 inhabitants in 2019) as it includes some of the most populous municipalities in the entire Metropolitan City.

It hosts a relevant infrastructural system, which includes the Turin-Milan high-speed railway line, the A5 and A4 motorways, the international airport of Turin-Caselle. It connects the city of Turin with the areas of Ciriacese to the west, Canavese to the north and Chivassese to the east.

At morphological level, the area is predominantly flat, characterized by a strong territorial homogeneity and by dependence relationships with the nearby chief town of Turin, of which it constitutes a northern hinterland area. A relevant aspect of the area is the relationship between the settlement fabric and the configuration of the local hydrographic network (made up of rivers and a system of historical channels, the so-called *bealere*), which exposes it to hydrogeological risks due to fluvial–torrential dynamics.

This relationship between the river and canal network and the high concentration of population has stimulated the experimentation of *ClimeApp* with reference to the flood risks for the built settlement, analysed according to the two IPCC emission scenarios RCP4.5 and RCP8.5 and with a medium, 2021–2050, and long-term time horizon, 2071–2100. Both scenarios (2021–2050 and 2071–2100) are compared with the period 1981–2010. RCP4.5, the *Strong Stabilisation* scenario, expects emissions to fall below current levels by 2070, while RCP8.5, the *Business-as-usual* scenario, assumes growth to continue at current rates, with an estimated increase by the end of the century of between 840 and 1120 ppm.

As in the HZP case study, the Climate Hazard indicators have been calculated by the CMCC Foundation within the EURO-CORDEX project.

The *Exposure* indicators used open-source data accessible through the regional geo-portal, the National Institute of Statistics (ISTAT) and data released by governmental agencies—e.g. those relating to the average taxable income per taxpayer. These indicators make it possible to model the presence of human activities with particular attention to settlement, production, areas classified as flood risk, services, and agricultural and cultural heritage assets. *Sensitivity* indicators, on the other hand, investigate phenomena such as the impact on the population of younger and older people, income, land consumption and the reduction of ecosystem services. *Adaptive Capacity* indices map the presence of services and initiatives that decrease vulnerability, such as the presence of hydraulic works, the level of digital connectivity ensured by ultra-wideband, the presence of institutions with volunteers, the involvement into institutional processes oriented towards adaptation and sustainability and the openness towards these issues.

9.4.2 Cross-Cutting Remarks and Indicators

The calculation of the indicators was carried out with GIS tools for all those indices related to areas such as hectares of residential buildings, agricultural and so on, or by counting, as in the case of the presence of hydraulic constructions. For the other indicators, operations were carried out through direct consultation of sources. In some cases, particularly regarding institutional indicators of *Adaptive Capacity* such as the level of openness to environmental sustainability issues, research was carried out to map the presence of institutions, health centres, universities or other legal forms whose very presence is indicative of attention to this issue. It should also be noted that for some indicators, the ISTAT data used dates to 2011, the year of the most recent census published by the Institute, which could differ significantly from the current demographic condition of the study area.

The indicators of *Exposure*, *Adaptive Capacity* and *Sensitivity* are used as a basis for the calculation of the risk together with the *Hazard* indicators, calculated according to the IPCC scenarios. Before describing the variation of the risk in the four scenarios, it is therefore appropriate to proceed with the description of the *Exposure*, the *Adaptive Capacity* and the *Sensitivity* that frame the situation of the HZNT in 2021. It should be noted that for the Municipality of Mappano, it was not possible to calculate numerous specific indicators as the Municipality had not yet been established at the date of acquisition of the necessary datasets. In any case, the values assumed by the neighbouring municipalities in the portion of the territory that these same municipalities have given to the new municipality can be referred to Mappano.

The *Hazard* indicators used to analyse the settlement system and floods framework relate to the increase in the frequency and intensity of rainfall with reference to the number of days of the year with rain, the cumulative and maximum daily precipitation over five consecutive days.[2]

The *Exposure* indicators considered are residential areas, industrial and commercial areas, supra-municipal services, historical and environmental assets, areas with agricultural use, areas with unrecognized use and the inhabitants settled in flood risk areas sanctioned by law (PAI areas). High exposure conditions are noted in Caselle Torinese, followed by Volpiano and Settimo Torinese.

Adaptive Capacity and *Sensitivity* define the dimension of *Vulnerability*. The *Adaptive Capacity* indicators considered are the presence of hydraulic infrastructures, institutional initiatives for adaptation to climate change, ultra-broadband coverage, institutions with volunteers and attention to environmental sustainability issues. Some of these indicators, especially the institutional ones, have been obtained with a qualitative survey of the plans and initiatives in force in the municipalities and at the supra-municipal level.

The *Sensitivity* indicators considered are, on the other hand, the incidence of buildings in a poor state of conservation, soil consumption, the reduction of ecosystem services in the woods, the population over 75 and children under the age of 6 and taxable income.

Overall *Vulnerability* is particularly high in the municipality of Borgaro Torinese, followed by Leinì and Settimo Torinese.

[2] The complete list is available in Ellena et al. (2020).

9.4.3 Results

The first risk scenario (RCP 4.5, 2021–2050) considers the IPPCC RCP4.5 emission scenario with a medium-term time horizon (Fig. 9.2a). The level of risk propensity compared to the current scenario for the settlement system with respect to the flood risk is higher in the municipalities of Settimo Torinese, Volpiano and Borgaro compared to the rest of the homogeneous area. The risk is related to a significant increase of *Hazard* in the eastern part of the territory, in particular in the three municipalities of San Benigno, Volpiano and Settimo Torinese.

The second scenario (RCP 4.5, 2071-2100), the simulation uses the same emission scenario but is projected over the period 2071–2100 (Fig. 9.2b). The most significant growth in risk compared to the current state is found in the three municipalities of Caselle Torinese, Leinì and Volpiano. The *Hazard* indicators in this case show a high

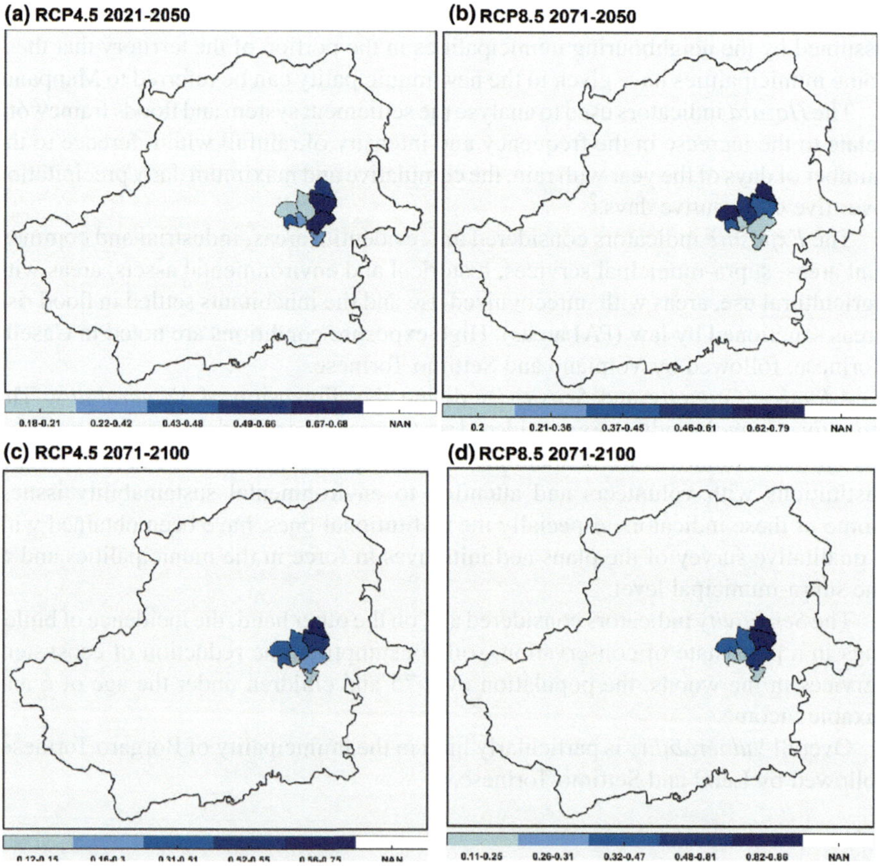

Fig. 9.2 From top left: Hazard scenarios: RCP4.5 2021–2050, RCP8.5 2071–2050, RCP4.5 2071–2100, RCP8.5 2071–2100

incidence in the north-northeast area of the territory, in particular in the municipalities of San Benigno and Volpiano followed by Leinì, Caselle and Borgaro.

The third scenario (RCP8.5, 2021–2050) uses the RCP8.5 emissions scenario, projected to the period 2021–2050 (Fig. 9.2c). The overall risk propensity with respect to flooding dynamics on the settlement system is increasing more markedly in the municipalities of Caselle Torinese and Leinì. In terms of *Hazard*, also in this case there is a higher incidence of indicators in the northern part of the area, with peaks in the municipalities of Caselle and San Benigno.

The fourth and last scenario (RCP8.5, 2071–2100), again adopting the RCP8.5 emission scenario, refers to the 2071–2100 timeframe and indicates a more marked propensity to risk in the entire central area, particularly in the Volpiano and Settimo Torinese territories (Fig. 9.2d). *Hazard* indicators related to flooding phenomena show more consistent growth in the whole northern and eastern area, in particular in the municipalities of San Benigno Canavese, Volpiano and Settimo Torinese.

9.4.4 Discussion and Conclusion

As the two case studies have shown, a tool such as *ClimeApp* can be a useful support for planning adaptation to climate change in areas where hydrogeological risks are relevant. As a matter of fact, *ClimeApp* allows to study how and where foreseen long-term changes in temperature and precipitation can increase these risks due to the increasing frequency and intensity of natural hazards. Moreover, the tool offers the opportunity to identify how much the risks are related to these hazards, or to features of the studied area such as its *Exposure*, *Sensitivity* and *Adaptive Capacity*. In these senses, the simulations carried out for the HZP and the HZNT have contributed to identify in which municipalities inside each of the two homogeneous zones is prior to promote adaptation; they also highlighted on which exposure and sensitivity factors it is important to focus actions in order to reduce the overall risks. At the same time, a comparison between the two—quite different—cases of application of the tool (Fig. 9.3) presented in this chapter let emerge some of its limits. Whereas the HZP is characterized by great heterogeneity between its components (three valleys, a wide plain, a hilly part, etc.) and the municipalities that make it up, the HZNT has a high level of internal homogeneity, also at a morphological level (Fig. 9.3). In the first case, it turns out to be difficult to compare the levels of hydrogeological risks among very different sub-areas inside the HZ; in the second case, conversely, the tool fails to identify consistent differences within the HZ. Probably, a level of heterogeneity intermediate between those of the HZP and the HZNT is the most appropriate for *ClimeApp*. A second limit is related to the relative—and not absolute—nature of the results that *ClimeApp* elaborates, due to the normalization processes that it uses. Indeed, *ClimeApp* specifies for each municipalities how great the level of risk is, not in absolute terms, but relatively to the levels of the others municipalities inside the HZ. If some of the indicators are changed for a municipality (e.g. simulating

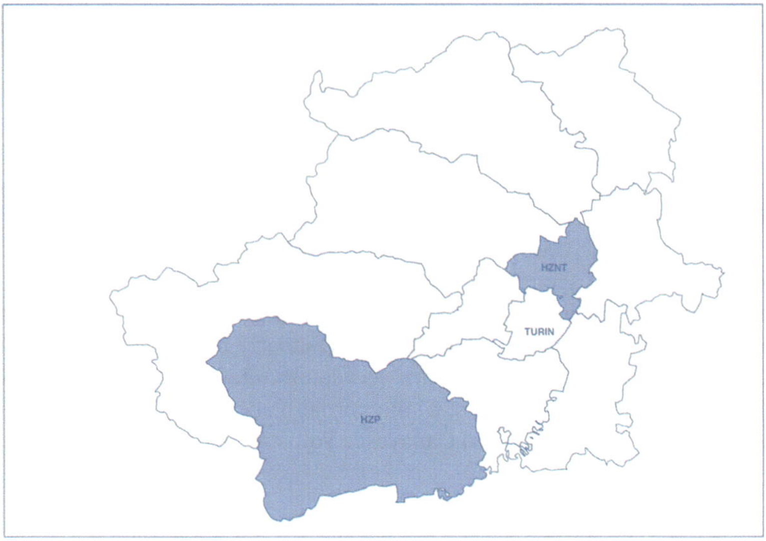

Fig. 9.3 Location of HZP and HZNT homogeneous zones in the territory of the Metropolitan City of Turin, Piedmont, Italy

some adaptation actions), *ClimeApp* shows how much the risks are reduced for this municipality in comparison with the risk levels of the others.

Considering these cautions, *ClimeApp* can support effectively public administrations in testing and defining adaptation strategies at an over-municipal scales, in particular at highlighting where and on which factors it is prior to focus actions and measures. This approach is particularly important for hydrogeological hazards and risks, which can be successfully faced only and just at a wide-area scale.

References

Birkmann J (2011) First-and second-order adaptation to natural hazards and extreme events in the context of climate change. Nat Hazards 58(2):811–840. https://doi.org/10.1007/s11069-011-9806-8

Bertoldi P (Ed) (2018) How to develop a Sustainable Energy and Climate Action Plan (SECAP) Part 2 – Baseline Emission Inventory (BEI) and Risk and Vulnerability Assessment (RVA). Publications Office of the European Union, Luxembourg. https://doi.org/10.2760/118857

Ellena M, Ricciardi G, Barbato G, Buffa A, Villani V, Mercugliano P (2020) Past and future hydrogeological risk assessment under climate change conditions over urban settlements and infrastructure systems: the case of a sub-regional area of Piedmont, Italy. Nat Hazards 102(1):275–305. https://doi.org/10.1007/s11069-020-03925-w

Field CB, Barros V, Stocker TF, Qin D (eds) (2012) IPCC (2012) Managing the risks of extreme events and disasters to advance climate change adaptation. Cambridge University Press, Cambridge, GB

GIZ, EURAC (2017) Risk supplement to the vulnerability sourcebook. Guidance on how to apply the vulnerability sourcebook's approach with the new IPCC AR5 concept of climate risk. GIZ, Bonn, DE

IPCC (2014) Climate Change 2014: mitigation of climate change. In: Edenhofer O, Pichs-Madruga R, Sokona Y, Farahani E, Kadner S, Seyboth K, Adler A, Baum I, Brunner S, Eickemeier P, Kriemann B, Savolainen J, Schlömer S, von Stechow C, Zwickel T, Minx JC (eds) Contribution of Working Group III to the fifth assessment report of the intergovernmental panel on climate change. Cambridge University Press, Cambridge, GB, and New York, NY, US

Oppenheimer M, Campos M, Warren R, Birkmann J, Luber G, O'Neill B, Takahashi K (2014) Emergent risks and key vulnerabilities. In: Field CB, Barros V, Stocker TF, Qin D, Dokken DJ, Ebi KL, Midgley PM (eds) Managing the risks of extreme events and disasters to advance climate change adaptation. Cambridge University Press, Cambridge, GB, and New York, NY, US, pp 1039–1099

Probst T, Wicki W, Zischg A, Pichler A (2013) Alpine strategy for adaptation to climate change in the field of natural hazards. Developed by the Platform on Natural Hazards of the Alpine Convention PLANALP. Platform on Natural Hazards of the Alpine Convention, Bern, CH

Schindelegger A, Kanonier A (2019) Natural hazard risk governance: Status Quo in the EUSALP Region. EUSALP Action Group 8

Zollner K (2018) United Nations platform for space-based information for disaster management and emergency response (UN-SPIDER). In: Brünner C, Königsberger G, Mayer H, Rinner A (eds) Satellite-based earth observation. Springer, Cham, DE. https://doi.org/10.1007/978-3-319-74805-4_24

Chapter 10
From Knowledge to Land-Use Planning: Local Resilient Experience in the Territory of the Municipality of Mappano

Luigi La Riccia and Angioletta Voghera

Abstract The chapter illustrates the result of the analysis on the municipality of Mappano, located in northern Turin (Italy). The studies were carried out as part of the collaboration between the Municipality of Mappano and the Inter-University Department of Territorial and Urban Studies and Planning (DIST) of the Politecnico di Torino, for the preparation of the first Municipal Urban Plan. The main goal was to contribute to this local planning tool by introducing innovative analyses, descriptions and elaborations which were useful in structuring planning choices. In particular, various data sources were systematized, integrated and coordinated to represent the territory from the point of view of both environmental phenomena and landscape in order to provide sustainability and resilience.

Keyword Landscape resilience · Mappano · Local ecological network · Viewshed analysis · Walkability · People-centred planning · Transformative resilience

10.1 A Landscape Resilience Perspective

Landscape resilience is "the process of transforming and designing the landscape to improve its quality, while also addressing adaptation and risk control needs (…) by putting the aspirations of the people at the centre" of the planning perspective (Voghera and Aimar 2022), and by bringing the environmental dimension into a dialogue with the landscape demands of the communities (EC, European Landscape

L. La Riccia (✉)
DIST—Interuniversity Department of Regional and Urban Studies and Planning, SDG11LAB, Politecnico di Torino, Turin, Italy
e-mail: luigi.lariccia@polito.it

A. Voghera
DIST—Interuniversity Department of Regional and Urban Studies and Planning, R3C—Interdepartmental Responsible Risk Resilience Centre, Politecnico di Torino, Turin, Italy
e-mail: angioletta.voghera@polito.it

G. Brunetta et al. (eds.), *Post Un-Lock*, The Urban Book Series,
https://doi.org/10.1007/978-3-031-33894-6_10

Convention, Florence, 2000). Indeed, landscape resilience in urban planning requires identifying "…priority places and modes for actions that foster new balances, with an attitude that has been appropriately called place-oriented and people-oriented" (Gabellini 2018, p. 96). Landscape resilience requires a place-centred approach and community empowerment, operating at the local scale and with attention to biodiversity and ecological networks, historical permanencies and perceptions. Looking forward, the approach to landscape resilience involves the maintenance and rehabilitation of cultural and environmental values (Winter et al. 2018), grounding conservation on "territorial governance" (Brunetta et al. 2019, p. 8) and social responsibility (Voghera 2015), by integrating landscape planning with local planning and design. Assumptions reaffirmed by the Peccioli Charter (2021), which establishes the "trans-scalar perspective" (Art. 10) of landscape resilience and the key role of "common territorial identity" (Art. 10).

Landscape planning and design must therefore be able to identify and to succeed in ensuring the quality of life, through an integrated landscape approach (Gambino and Peano 2015), which is crucial for interpreting the territory holistically, from a socio-economic, environmental and perceptual perspective, creating synergies between adaptation, and landscape and biodiversity policies.

10.2 A Case Study of Local Resilience

The community of Mappano has recently been recognized as a municipality by the Piedmont Region (Law No. 1/2013, entitled "Establishment of the Municipality of Mappano"; B.U.R. Piedmont—No. 5 of 31.01.2013). Since this is the urban plan of a newly formed municipality, the analyses carried out call for a historical reconstruction of the transformations of the territory with particular attention to the process of formation of the village settlement of Mappano as an unplanned sedimentation of fragmented and contradictory settlement choices. In fact, this aspect can be traced to not only by the effects of the diverse urban planning tools of the neighbouring municipalities of Caselle Torinese, Borgaro Torinese, Leinì and Settimo Torinese but also by the population's aspirations and community identity (Traore 2021).

Considering these aspects, Mappano cannot be interpreted, for the purpose of drafting the first urban plan, within the limits of its administrative boundaries, but in its location in a complex node (Lanzo and Canavese Valleys, the urban metropolitan area along the Stura River). Besides, Mappano is an environmental area of great biodiversity and landscape richness for the green and blue infrastructure policy at metropolitan scale, considering the Corona Verde Project (Green Crown) and the Tangenziale Verde Strategy of the General Metropolitan Regional Plan of Turin (PGTM).

Considering the community aspirations, Mappano is characterized to be redefined in terms of socio-economic, cultural possibilities and public services (structure of the population, intergenerational relations, the structure of the family, school, leisure, education, location of sports centres). The housing system is weak and with limited quality and needs diversification of the inhabited area (sub-services) and agricultural systems in relation to their quality and maintenance and visual perceptual aspects.

Why is Mappano an interesting case study? The case study of Mappano is emblematic in demonstrating the role played by the supremacy of local identity or local interests despite the recognized importance of the key role played by land everywhere (Pileri and Scalenghe 2016). The contradiction highlighted by this case raises the discussion of some crucial issues related to the role of local urban planning and soil protection, which cannot be fragmented or subject to short-term local interests. Mappano experienced strong demographic and economic growth in line with other municipalities in north-western Italy participating in the broader process of suburbanization of Turin during the '80s. The availability of large open areas allowed for the programming of important land conservation linked to the creation of a large local and broader regional park that extends from the towns of Borgaro and Settimo and is part of the larger project of the Corona Verde of the metropolitan area of Turin. At the same time, various infrastructure interventions and urban areas regeneration were programmed. This area is located nearby to the strategic infrastructures serving the entire Turin metropolitan area: the A4 motorway, the airport of Caselle, the high-speed railway to Milan and the Canavesana railway line. In addition, the construction of line 2 of the Turin subway towards Settimo Torinese and the valorization of the Turin–Ceres railway, connecting the airport to Turin's city centre, are planned.

At the COVID-19 time, Mappano redefines its role within the first town plan process overcoming the crisis; this process requires reflection on the interpretation of the territory as a synthesis of the relationship between community and environment.

The following paragraphs describe the elaborations carried out in the framework of the collaboration agreement signed between the Politecnico di Torino[1] (DIST and DIATI Departments and the Responsible Risk Resilience R3C Interdepartmental Centre) and the Municipality of Mappano, which is aimed at strengthening the definition of studies for the urban plan role through a multidisciplinary approach for sustainable and resilience development.

10.3 Analysing and Understanding the Environmental and Landscape System

Attributing an ecological and ecosystemic role to the territory means considering a general renewal of urban planning paradigms, taking into account the importance of rural production and entrepreneurial and political interests. In line with these objectives, the new ecological corridors should not be "spatial schemes", perhaps excellent in aesthetic terms, but lacking in biodiversity operability (Voghera and La

[1] Departmental agreement among Politecnico di Torino (Departments: DIATI and DIST and R3C) and the Municipality of Mappano (2021–2023), responsible: Lingua A. and Voghera A. Research group: DIST-R3C: Grazia Brunetta, Ombretta Caldarice, Luigi La Riccia, Ammj Traore, Giulia Matteucci, Mattia Scalas; DIATI: Stefano Angeli, Valeria De Ruvo, Paolo Maschio, Marco Piras.

Riccia 2019). For this reason, when investigating the state of naturalness and diversity at different scales, it is necessary to go further in order to prioritize the proposal of ecological coherence: to consider the network in the context of the impacts of human activities and, more generally, in urban planning processes with reference to their operability (La Riccia 2015b).

In this context, several interesting studies on this topic have been launched in the Piedmont region (Italy), with the aim of improving the overall ecological quality of natural and landscape areas and specifically indicating how to avoid ecological fragmentation and strengthen biodiversity (Voghera and La Riccia 2016, 2019). In recent years, specific research has been conducted by the Polytechnic of Turin in collaboration with the Metropolitan City of Turin and ENEA with the aim of defining a proposal for the implementation of the ecological network at a local level in some areas of the territory.[2]

In the last two years, other experimentations have been conducted in other municipalities (Mappano, Alpignano and Moncalieri) following the developed methodology, adapting it to different geographical contexts and taking into account new perspectives in relation to post-pandemic needs. In Italy, the PNRR (Italian Next Generation EU Plan, 2021) is contributing to pushing local governments to prepare new projects for the country's economic recovery. The goal is the ecological transition but also digitization, competitiveness, training and social, territorial and gender inclusion.[3]

In line with PNRR, the proposed approach was to guide local governments with specific measures to limit anthropogenic land use and, where possible, to promote the conservation of ecosystem services. Habitats, natural areas and landscapes were not only interpreted from an exclusively ecological point of view (a mosaic of ecosystems) but also considering a broader perspective that embraces the cultural, social and economic values of the territory. The proposed methodology identifies the ecological character of the territory and defines criteria for the evaluation of the different land-use types: in Piedmont, 97 land-use types were identified, according to the Corine Land Cover database. Subsequently, six key indicators (Voghera and La Riccia 2016, 2019) were applied to assess ecological status (see Fig. 10.1):

- *Naturalness*: land-use types are classified into five levels of naturalness, considering the proximity to formations that would be present in the absence of disturbance (climax). Thus, the levels of naturalness range from level 1, which includes all natural formations, to a maximum of level 4, which considers land-use types

[2] Between 2014 and 2016 the research "Guidelines for the Green System of PTC2" (convention between Metropolitan City of Turin, ENEA and Polytechnic of Turin) and the "Operational proposals for the ecological network of Chieri" (Polytechnic of Turin and Comune di Chieri, Turin) were conducted with the objective of defining a proposal for the implementation of the ecological network at the local level firstly in two municipalities of Turin (Ivrea and Chieri).

[3] The six major areas of intervention (pillars) on which the PNRR focuses are: (1) Green Transition, (2) Digital Transformation, (3) Smart, Sustainable and Inclusive Growth, (4) Social and Territorial Cohesion, (5) Health and Economic, Social and Institutional Resilience, (6) Policies for the new generations, children and young people.

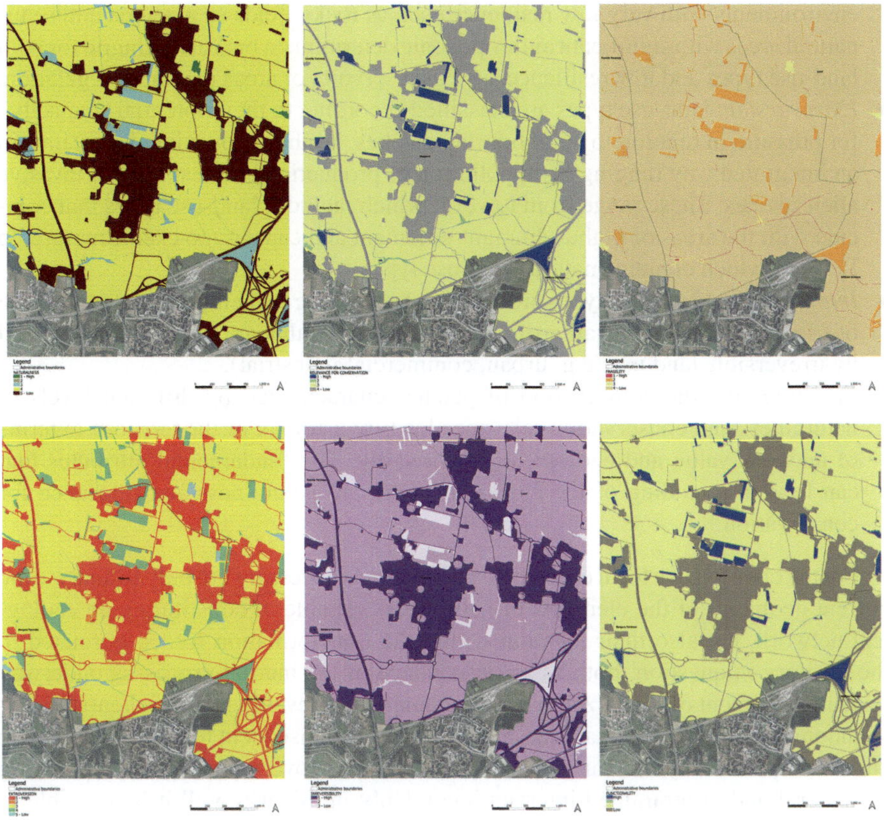

Fig. 10.1 Maps of Mappano territory according to the considered indicators (in the order of appearance): naturalness, relevance for the conservation, fragility, extroversion, irreversibility, functionality. *Source* Voghera and La Riccia

that are fully anthropogenically determined but not artificial (such as almost all cultivated land) and level 5, which includes land-use types corresponding to artificial areas.

- *Relevance for conservation*: land-use types are classified into four levels of relevance according to the relevance/suitability of the land use for the conservation of biodiversity, while considering the importance for habitats and species. The concept of habitats of interest for Natura 2000 Network species is introduced, which includes not only habitats of Community interest, but also complex habitats whose conservation is necessary for the protection of Natura 2000 Network species.
- *Fragility*: land-use types are classified in terms of their intrinsic fragility due to pressures such as pollution, exotic and invasive species entry, and anthropogenic disturbance in general. Level 1 includes land-use types that define both natural

environments with very low resilience such as rocky fields or glaciers, and semi-natural areas with significant anthropogenic determinism but easily fragile for both land-use types and low resilience such as reservoirs or areas of sparse vegetation.

- *Extroversion*: land-use types are classified according to their potential "capacity" for pressure in relation to neighbouring patches. We have considered pressures in an integrated way ranging from pollution of production to the spread of invasive alien species. These range from Level 1, which includes land-use types that coincide with the areas of highest human settlement and capacity to exert pressure, to Level 5, which contains natural land-use types.
- *Irreversibility*: land-use types are classified according to their potential for change of use. Level 1 includes all artificial land-use types that are totally characterized by irreversible land use (e.g. urban, commercial industrial).
- *Functionality*: the combination of patches characterized by different levels of naturalness and conservation relevance leads to a zoning of the territory in terms of network value and ecological functionality. The fundamental attributes that can lead to a reading of the current network are naturalness and relevance for conservation.

For the creation of a local ecological network, it is necessary not only to analyse the current geometry of the elements of naturalness capable of constituting an ecological network, but also their location within the transformation forecasts related to the territory in question, both as a consequence of the inertial processes underway (e.g. advancement of urbanization fronts, changes in prevailing crops, phenomena of abandonment of hill–mountainous areas), and those consequent to the programmatic choices expressed by the various levels of territorial government (general planning or sectoral and programmed interventions). Only in this way will it be possible to prefigure an overall design of the ecological network capable of achieving the set objectives, demonstrating compatibility with the objectives of the various sectors.

From the integration of the results of different indicators, the so-called structural map of the ecological network (Voghera and La Riccia 2016, 2019) has been obtained (see Fig. 10.2). This map shows the elements of the local ecological network system, chosen on the basis of the levels of naturalness, ecological functionality, geographical continuity and consists of three main elements:

- *Structural elements of the network* (primary ecological network), i.e. areas with high and moderate ecological functions, as well as areas hosting specific conservation emergencies, i.e. of natural and significant importance for the conservation of biodiversity.
- *Priority Network Expansion areas*, i.e. areas with a residual ecological function in which action to increase the functionality of the primary ecological network is a priority and for which the implementation of protection measures to maintain the primary ecological network is planned. These areas are further subdivided into connection areas and portions contiguous to structural elements.
- *Possible expansion of the network areas*, i.e. areas with residual ecological functionality, but on which it is possible to implement new measures aimed at

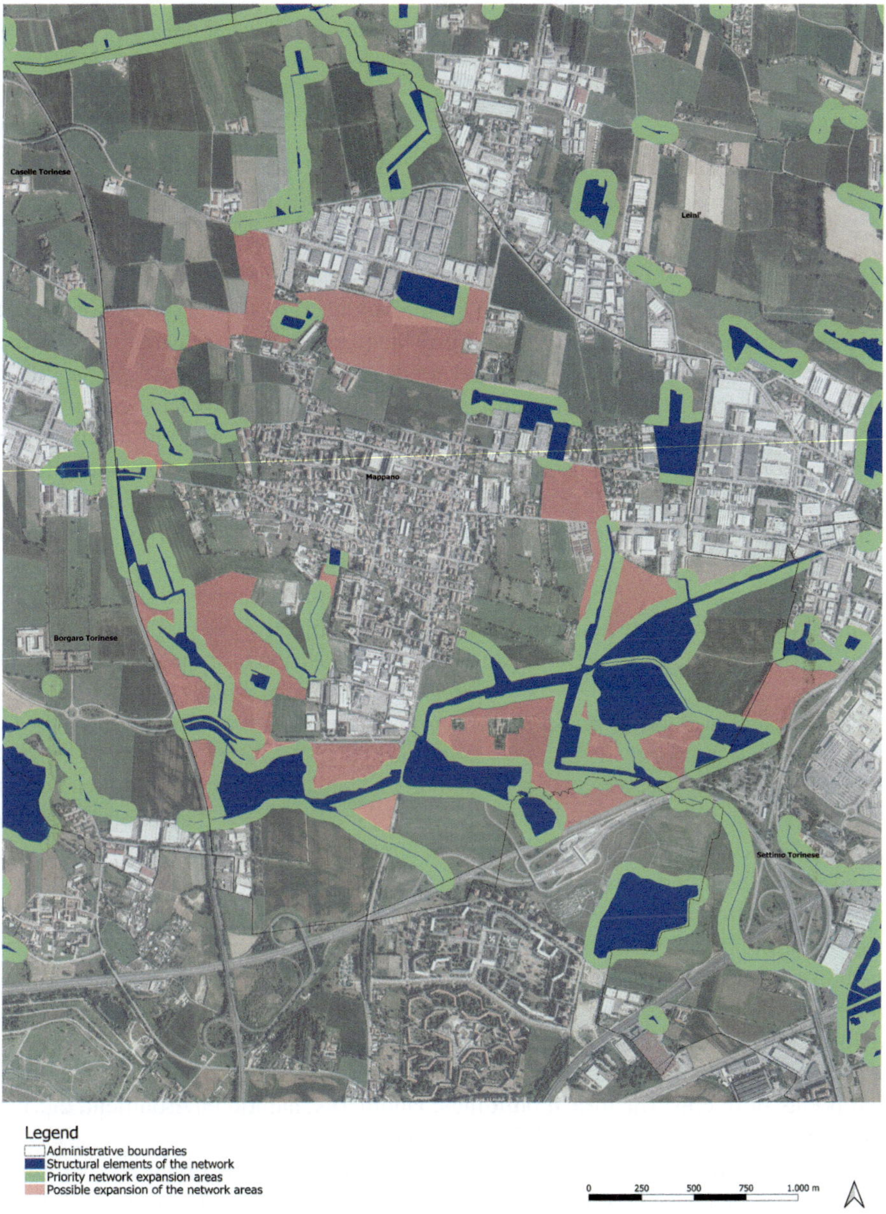

Fig. 10.2 Map of the ecological structurality of Mappano territory. The picture shows the two components of ecological structurality (structural elements and the contiguous portions to the structural elements). *Source* Voghera and La Riccia

increasing naturalness that are useful for protecting the habitat and species of interest for the conservation of biodiversity.

10.4 Analysing and Understanding the Environmental and Landscape System: Landscape Sensitivity

With the advent of ICT, the wide use of geospatial data and the creation of Digital Elevation Models (DEM) and Digital Surface Models (DSM), the development and implementation of new GIS methodologies help to determine visible areas more accurately and automatically (Travis et al. 1975; Yoeli 1985; La Riccia 2015a, 2017; Chiesa and La Riccia 2016). The family of GIS software can provide a spatial representation of landscape elements, studying the intervisibility relationship between different points more or less distant from each other, and define the overall sensitivity of the landscape. The aim of this analysis is to contribute to the field of urban planning, taking into consideration the objective conditions and geometries of different points of analysis (formal characteristics of the landscape scene, observation points, radius and depth of view, perceptual reference points), assuming that they can be predictive of a subjective landscape experience, a subjective perception. The viewshed analysis simulates the relationship between landscape morphology and spatial elements and helps to calculate the coverage (visual space) with respect to the position and visual horizon of a specific observer. Based on a model (DTM or DSM), it is possible to perform the analysis from individual positions (viewsheds), paths (incremental viewsheds) or areas (cumulative viewsheds). In all cases, the viewshed defines the assumed view space as the portion of the landscape that an observer can see. This process is not only based on the three-dimensional aspects of the space but also on other conditions such as the position of the observer (altitude, proximity, etc.), the direction of the view (azimuthal and vertical angles) and atmospheric conditions (minimum and maximum visibility radius). The results are based on a Boolean visibility concept and are given in binary code (1 = visible; 0 = not visible). A binary viewshed answers a fundamental question: What portion of the landscape is visible from a given vantage point? When carrying out this analysis, it is important to include all kinds of information about, for example, other scenic elements or particular points of interest (historical buildings, landmarks, natural environment, etc.) in order to assess different intervisibility relationships.

The geometric characteristics of each selected scene are organized within a geographical database that includes various elements: altitude of the selected viewpoint, height of the observer relative to the ground, height of a visual landmark, width of horizontal and vertical angles, depth (radius) of the view horizon. The set of parameters (La Riccia 2017) that can be imported by the GIS software is shown below:

- Spatial coordinates of the viewpoint;
- SPOT: altitude of the viewpoint;

- OFFSET A: height of the observer with respect to the ground;
- OFFSET B: height of a different landmark or another point of interest;
- AZIMUTH 1 and 2: width of the horizontal angle;
- VERT 1 and 2: width of the vertical angle;
- RADIUS 1 and 2: minimum and maximum distances (radius) of the view.

When several viewshed analyses are obtained from several points, it may be possible to superimpose them and create an "absolute visibility" map of the landscape. The result can be a Boolean (raster) image, or even be characterized by a more complex subdivision by incorporating the different viewshed analyses into a single map, generated by superimposing several raster images through the "combine" function of the GIS, and taken as the result of the "landscape sensitivity" index.

Figure 10.3 shows 6 classes of landscape sensitivity and derives from the weighted sum of the individual viewshed analyses, calculated from each of the seven viewpoints selected on Mappano following on-site inspections. The methodology therefore makes it possible to contribute to the drafting of landscape protection indications that can be applied differently, protecting the views on the basis of three distinct levels (foreground, medium-ground, background) or defining specific management plans that include detailed indications, for example the coherence between views, the development of the local ecological network and urban transformations, the requests for environmental mitigation and compensation to overcome the most critical territorial situations.

10.5 Analysing and Understanding the Environmental and Landscape System: Accessibility and Sustainable Mobility

The walkability assessment in Mappano was conducted through the improvement of an analysis methodology that aims to recognize the parts of the city where the actions for the improvement of walkability could be more effective.

The concept of walkability is a way of looking beyond the presence, distribution and simple accessibility of urban facilities (Cittadino et al. 2020; La Riccia et al. 2021): the spatial quality and the ability to accommodate and facilitate pedestrian mobility in the urban environment influence the way in which people perceive and use the city. What the concept of walkability brings with it, in fact, is the quality of accessibility: how and to what extent the urban environment is able to promote walking and offer itself as a platform for a daily life based on pedestrian mobility. But the research we have analysed often describes a path that has led to the construction of walkability indices, taking into consideration density and the urban mix (which brings together possible points of origin and destination of movements), safety (which concerns both intersections between pedestrian paths and vehicular paths, which anthropogenic safety), the pleasantness of the environment (quality of pavements,

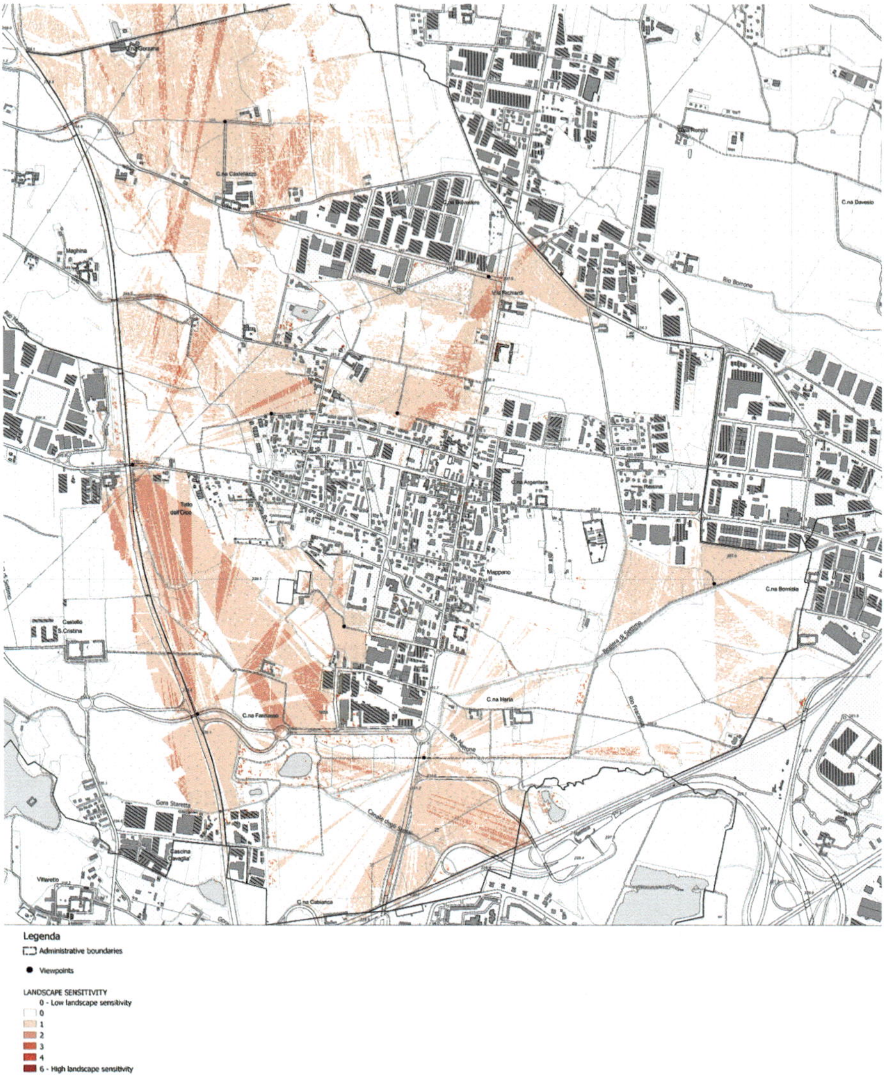

Fig. 10.3 Landscape sensitivity map from predetermined observation points (cumulative viewshed). *Source* Voghera and La Riccia

presence of shops and other activities along pedestrian paths, presence of greenery, low level of pollution and noise, etc.).

An analysis of accessibility is almost always present, expressed as the distance to be covered on foot to reach certain services, which however is calculated on the road network, that is, using the data and tools developed for vehicular traffic. This is obviously due to the unavailability of pedestrian path graphs. But this introduces significant distortions. In Mappano, however, the "walkable" space was modelled through a raster with a resolution of 1 m × 1 m: each cell of the raster was assigned an "impedance to be walked" value (cost raster, "local cost" of QGIS). The cost raster was then used to calculate the accessibility to some walking mobility destinations, such as the cumulative weighted distance (cost distance).

With this goal in mind, criteria were focused and indexes were constructed to make them operable. All data was checked and rasterized with 1 × 1 m cells. The other data, of a point nature, have been spatialized using the kernel density estimation (KDE) which, given the value of a phenomenon in a point, represents its diffusion and attenuation in a circular neighbourhood, with a radius suitably defined in relation to the phenomenon represented.

All the maps were produced as rasters with cells of equal size: the different rasters were then simply added up with the Map Algebra method, giving each a suitable weight (Table 10.1; Fig. 10.4). The use of the weighted sum, avoiding more complex algorithms, is functional to maintain a certain control over the meaning of the results, very appropriate because these results derive from a procedure of a certain complexity, which contains several critical steps inside. In identifying the criteria at the neighbourhood level, the starting point was the "Walkability Hierarchy of Needs Pyramid" (ITDP 2018, p. 13–14). The six proposed criteria have been compacted into three: practicability, safety (physical safety and anthropogenic safety) and comfort/pleasantness.

The weighted sum of the macro-indices relating to these three criteria constitutes the cost raster, readable as a detailed representation of the walkability (Fig. 10.5). High values of the macro-indices mean high practicability, high safety, high pleasantness, while the values of the raster cost represent a cost, an impedance to walk the cell. The values of the cost raster were then calculated as a complement to 100 of the normalized sum of the macro-indices.

Improving walkability therefore means intervening on those extrinsic characteristics to people that favour full individual expression. Our research therefore proposes a two-level reasoning: walkability has been assumed as a complex indicator for assessing the state of places (at the neighbourhood level) and becomes a tool for orienting the design action aimed at improving livability (at the city level), with the aim of recognizing the parts where actions aimed at improving walkability can be most effective.

10.6 Conclusion: Reasoning in an Integrated Perspective

The set of analyses produced based on consolidated methodologies involves rules, models and criteria for solving common problems for planning requiring reasoning in an integrated perspective.

The work carried out is based on the idea of outlining a shared theoretical perspective, because it is based on theoretical acquisitions already given, but which required measuring oneself with the territorial dimension of Mappano to guide them towards new hypotheses, methodologies and application directions. By dealing with urban

Table 10.1 Set of indicators (at city level and at neighbourhood level) for evaluating walkability in Mappano

Macro-indexes		Indicators	
Weight attributed in the weighed sum (%)	Description	Weight attributed in the weighed sum (%)	Description
Walkability at city level			
40	Population density	30	Density of children 0–14 years old, inhab/ha, by census sections
		50	Population density 15–64 years old, inhab/ha, by census sections
		20	Population density over 65 years old, inhab/ha, for census sections
20	Density of economic activities	60	Local unit density, LU/ha, by census section
		40	Density of employees, employees/ha, by census section
20	Urban form	50	Block density per sqm/km, by statistical area
		50	Density of public pedestrian circulation areas (by census section) sqm/sqm
20	Attractive activities for services, leisure, intermodality	30	Number of Public Facilities (e.g. security, administrative offices, social welfare services)
		20	Number of Schools (Schools, Kindergartens, Universities)
		20	Number of leisure places (cinemas, sports facilities, thematic markets, museums, tourism offices, theatres)
		30	Number of Intermodal nodes (taxi stations, car sharing, bike sharing, stalls of disabled people, parking areas, metro, bus and train stations and stops)
Walkability at neigh-bourhood level			
40	Accessibility	50	Presence of sidewalks, pedestrian crossings, paths, slides, stairs…
		50	Population density 15–64 years old, inhab/ha, by census sections
20	Security	10	Presence of activities at the sidewalk level: shops, stalls
		20	Presence of intersections with vehicular traffic regulated by traffic lights
		20	Separation of pedestrian/vehicular routes
		10	Number of accidents involving pedestrians

(continued)

Table 10.1 (continued)

Macro-indexes		Indicators	
Weight attributed in the weighed sum (%)	Description	Weight attributed in the weighed sum (%)	Description
		15	Presence of areas and intersections with controlled vehicular traffic
		20	Adequate lighting levels
		5	Social control by the people of the houses overlooking the lower floors
20	Comfort/ pleasure	15	Presence of trees
		5	Presence of street furniture
		10	Presence of attractive activities at the sidewalk level (shops)
		5	Presence of covered paths
		5	Presence of water points
		15	Sidewalk quality
		5	Low noise level
		20	Path contiguous to green areas
		5	Presence of "buildings of particular historical interest"
		5	Presence of works of art
		10	Presence of landscape visuals and panoramic points

Source Voghera and La Riccia

planning, in particular, such an approach helps us to abandon the idea of relegating this activity within a single disciplinary area and to acquire the conviction of taking a look "beyond the borders", under penalty of reductivism and, consequently, of not being able to grasp the relevant threats on the territory due to the global changes in progress.

A people-centred and place-based approach for urban planning that starts from ecological and landscape perspective can contribute to make operational the concept of transformative resilience, proposing solutions for the ecological and landscape quality needed for the coevolution of the territories (Folke 2016). The Mappano experience enforces the community aspiration to construct new development path based on shared visions about landscape and nature preservation, making effective the "nature-based recovery" or "nature positive economy" (IUCN Marsiglia 2020) towards a landscape resilience.

Fig. 10.4 Walkability at city level in Mappano. *Source* Voghera and La Riccia

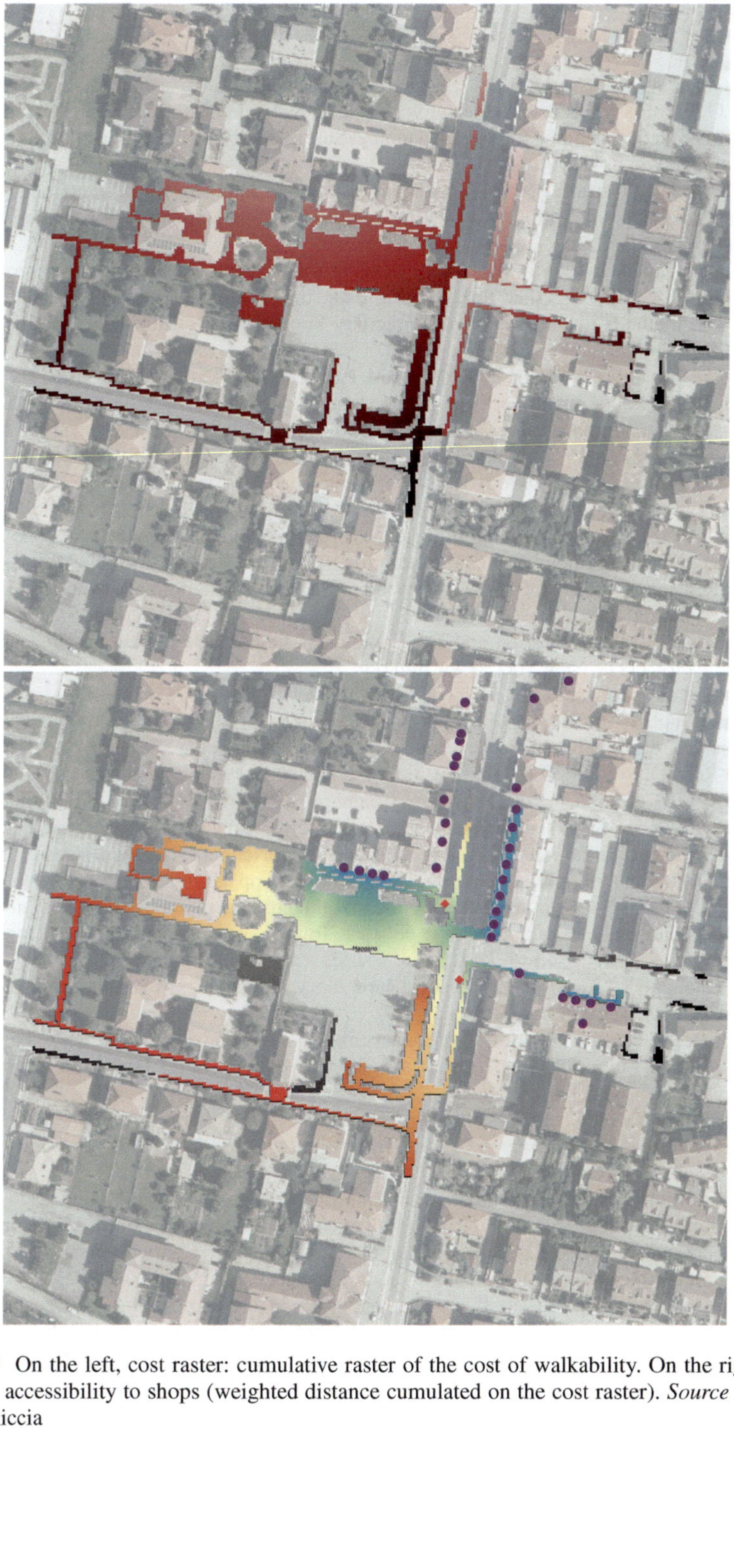

Fig. 10.5 On the left, cost raster: cumulative raster of the cost of walkability. On the right, cost distance: accessibility to shops (weighted distance cumulated on the cost raster). *Source* Voghera and La Riccia

References

Brunetta G, Ceravolo R, Barbieri CA, Borghini A, de Carlo F, Mela A, Beltramo S, Longhi A, De Lucia G, Ferraris S, Pezzoli A, Quagliolo C, Salata S, Voghera A (2019) Territorial resilience: toward a proactive meaning for spatial planning. Sustainability 11:2286. https://doi.org/10.3390/su11082286

Charter P (2021) La Carta di Peccioli. La nuova costituzione della nazione delle comunità resilienti italiane. Available via: https://www.comunitaresilienti.com/centro-di-ricerca-sulla-resilienza/. Last access 29/05/2023

Chiesa G, La Riccia L (2016) Tools and methods for evaluating and designing the perceived landscape. 3D-GIS, viewshed analysis, big data. In: Pagani R, Chiesa G (eds) Urban data. Tools and methods towards the algorithmic city Franco Angeli, Milano, pp 159–198. ISBN: 9788891738646

Cittadino A, Garnero G, Guerreschi P, Eynard E, Melis G, Fiermonte F, La Riccia L (2020) Improving the walkability for next-generation cities and territories, through the reuse of available data and raster analyses. In: Tira M, Pezzagno M, Richiedei A (eds) Pedestrian, urban spaces and health. CRC Press—Taylor and Francis Group, Leiden, pp 177–181. ISBN: 9781003027379

Folke C (2016) Resilience, Oxford Research Encyclopedia of Environmental Science. Oxford University Press, pp 1–63. https://doi.org/10.1093/acrefore/9780199389414.013.8

Gabellini P (2018) Le mutazioni dell'urbanistica. Principi, tecniche, competenze. [The mutations of urban planning. Principles, techniques, skills.]. Carocci. ISBN: 978-8-84309-366-3

Gambino R, Peano A (eds) (2015) Nature policies and landscape policies. Towards an alliance. Springer, Dordrecht

ITDP, Institute for Transportation and Development Policy (2018) Pedestrians first, tools for a walkable city. ITDP, New York. www.itdp.org

La Riccia L (2017) Landscape planning at the local level. Springer, Cham

La Riccia L, Voghera A, Salizzoni E, Negrini G, Maltoni S (2021) Planning ecological corridors: a cost distance method based on ecosystem service evaluation in the Sardinian Cork Oak Forests. In: La Rosa D, Privitera R (eds) Innovation in urban and regional planning. Proceedings of the 11th INPUT Conference, vol 1. Springer, Cham, pp. 301–310. ISBN: 978-3-030-68823-3

La Riccia L (2015a) Sistemi informativi geografici per la valutazione dell'impatto scenico dei paesaggi industriali. In: Aliamo A, Aru S, Donadelli G, Nebbia F (a cura di) Geografie di oggi. Metodi e strategie tra ricerca e didattica. Franco Angeli, pp 41–56. ISBN: 9788891724434

La Riccia L (2015b) Nature conservation in the urban landscape planning. In: Gambino R, Peano A (eds) Nature policies and landscape policies. Towards an alliance. Springer, Dordrecht, pp 157–164

Pileri P, Scalenghe R (2016) The impractical supremacy of local identity on the worthless soils of Mappano. City, Territ Archit 3(5)

Traore A (2021) Geospatial analysis for urban cycling planning and design. Master's Thesis. Supervisor: Andrea Lingua. Co-supervisors: Angioletta Voghera, Luigi La Riccia. Politecnico di Torino

Travis MR, Elsner GH, Iverson WD, Johnson CG (1975) VIEWIT: computation of seen areas, slope, and aspect for land-use planning, General Technical Report PSW-11/1975, Forest Service, U.S. Department of Agriculture

Voghera A, Aimar F (2022) Towards a definition of landscape resilience: the proactive role of communities in reinforcing the intrinsic resilience of landscapes. In: Carta M, Perbellini MR, Lara-Hernandez JA (eds) Resilient communities and the Peccioli Charter. Towards the possibility of an Italian Charter for resilient communities. Springer, Cham, pp 111–120

Voghera A, La Riccia L (2019) Ecological networks in urban planning: between theoretical approaches and operational measures. In: Calabrò F, Della Spina L, Bevilacqua C (eds) New metropolitan perspectives. Local knowledge and innovation dynamics towards territory attractiveness through the implementation of horizon/E2020/Agenda2030, Springer, Cham, pp 672–680. ISBN: 978-3-319-92101-3

Voghera A, La Riccia L (2016) Landscape and ecological networks: towards a new vision of sustainable urban and regional development. LaborEst 12:89–93

Voghera A (2015) Resilience through community landscape project. In: D'Onofrio R, Sargolini M (eds) Resilient landscapes for cities of the future—UNISCAPE En-route international seminar. University of Camerino, pp 103–108. ISSN: 2281-3195

Winter KB, Lincoln NK, Berkes F (2018) The social-ecological keystone concept: a quantifiable metaphor for understanding the structure, function, and resilience of a biocultural system. Sustainability 10:3294. https://doi.org/10.3390/su10093294

Yoeli P (1985) The making of intervisibility maps with computer and plotter. Cartographica 22:88–103

Chapter 11
Space for Rights. The School Between Planning Standard and Social Innovation

Daniela Ciaffi, Carolina Giaimo, Emanuela Saporito, and Valeria Vitulano

Abstract The paper straddles the boundary between urbanism and sociology, working on the common ground of rights: public services and facilities ('planning standard,' according to the Italian spatial planning legislation) on the one hand and the care of the commons on the other, in addition to attempting to grapple with a third dimension of rights that places space and society alongside law. This methodological hypothesis is practiced from the critical analysis of one of the basic public facilities: the school. In the national debate on public services and common goods, school spaces are one of the recurring examples of how they have functionality as public educational services during school time and how they can also have other functional profiles as common goods, i.e., as civic centers open to the urban community during out-of-school hours. The theme of the hybridization of spaces and functions emerges with ever-increasing theoretical and empirical force in the reflections on so-called social innovation. Even if, in several cases, many people ignored one of the most beautiful definitions of planning standards by Giovanni Astengo (1966). He stated that, besides being a minimum, standards represent a minimum of civilization. The paper intends to bring attention to the complex value of spatial and social resources related to schools.

Keywords Urban facilities · Educational services · Social innovation

---

D. Ciaffi · E. Saporito · V. Vitulano
DIST—Interuniversity Department of Regional and Urban Studies and Planning, Politecnico di Torino, Turin, Italy
e-mail: daniela.ciaffi@polito.it

E. Saporito
e-mail: emanuela.saporito@polito.it

V. Vitulano
e-mail: valeria.vitulano@polito.it

C. Giaimo (✉)
R3C—Risk Resilience Centre, DIST—Interuniversity Department of Regional and Urban Studies and Planning, Politecnico di Torino, Turin, Italy
e-mail: carolina.giaimo@polito.it

G. Brunetta et al. (eds.), *Post Un-Lock*, The Urban Book Series,
https://doi.org/10.1007/978-3-031-33894-6_11

11.1 Introduction

11.1.1 Planning Standards, Urban Welfare, and the Spatial Structure of the City

The responsibilities of urbanism concerning the inequalities that manifest in cities are intrinsic to the principle on which planning activity has historically been based, namely the functional division of space through zoning. The control of space organization through planning determines perimeters of distinction and exclusion that have evident and explicit effects on the control of the social organization, showing the complex relationship between space and power (Mazza 2015). The recent pandemic has highlighted the problematic condition experienced by urban society in the contemporary city where social exclusion and inequalities are growing, even though the equal social dignity of all citizens is a fundamental principle of the Italian constitutional system. For some time now a new urban issue has been emerging that underlies generalized conditions of marginality and which, also as a result of the dramatic situation of the global health emergency, calls for wide-ranging glances and new considerations and interpretations for the activation of a new urban welfare, aimed at guaranteeing all local communities the right to health, housing, education, environment, public mobility, and the city.

In the face of such evidence, while it is true that town planning cannot be blamed for eliminating social inequalities, there is also no doubt that it can be entrusted with the task of defining settlement scenarios that can offer tendentially equipotential access to welfare spaces, to those essential public services, of collective interest, that represent the structuring and vital element of the urban system. Moreover, the 'trauma' generated by the pandemic and how it has called into question the space–time relationship in human relations, in the forms of production and living, and the spatial rules of the city's composition asks urban planning to question the ways and forms of innovation of its action, returning to investigate the spatial structure of the city starting from public space, as a space of guarantee of citizens' fundamental rights (Giaimo 2021). Urban planning must therefore question itself on the ways and forms of innovation of its action, returning to investigate the spatial structure of the city starting from public space as a space to guarantee the fundamental rights of citizens.

For this reason, urban regeneration policies of many cities are moving toward the multifunctional enhancement of the school facilities network that represents the main proximity structures that can perform strategic welfare functions (Giaimo 2022).

11.1.2 Common Goods and Social Innovation

We know that when we talk about urban regeneration we mean both physical and social transformations. But much less widespread is the awareness that talking about urban regeneration means talking about the right to take care for common resources that we all should have: really being able to—and not having to—take care of both spaces and services, people and even animal populations and plants that inhabit and live in cities and territories. It is important to note that Italy is an exceptional laboratory concerning this new 'right to care' (Ciaffi 2019) at the international level, thanks to a series of real revolutions in administrative, public, and private law which started from the beginning of the present century have made our jurisprudence at the forefront (Arena and Bombardelli 2022). In 2001, the *Principle of Horizontal Subsidiarity* was introduced in our Constitution stating that everyone can cooperate with public institutions to carry out actions of general interest (art. 118 last paragraph). Under this constitutional umbrella, from 2014 to today, about three hundred Italian municipalities have adopted the *Regulation for the shared administration of common goods* which allows the stipulation of collaboration agreements between active citizens and local public administrators. The national *Subsidiarity Laboratory* (Labsus) periodically draws up its Reports to photograph the progress of the thousands of *Pacts of collaboration* through which different subjects come together in smart alliances for purposes of general interest. There are tens of thousands of Italians who, more or less consciously, have become aware of their right to take care of common goods together, in the last decade, starting very different and creative spatial and immaterial actions. In the majority of cases, these are actions for the regeneration of the city and the territory: national data on 2021 tell us that the commons goods focused by the communities in action are the environment and urban greenery (in 48% of cases), urban furniture (19%), cultural heritage and initiatives (7%), schools and educational initiatives (7%), social inclusion (7%), co-planning of a series of interventions (4%), sports (3%), intangible common goods (2.5%), animation of the territory (1.7%), and other (Labsu 2022).

What happens in the meantime at a disciplinary level? Urban planning is obviously a discipline known for its at least partly law-related nature: Every planner or architect is confronted with laws and regulations. On the contrary, environmental and social psychology experts, city and territory sociologists, anthropologists, and ethnographers, who also often embrace participatory approaches, rarely start from the right to the city (Lefebvre 1970) also understood as knowledge of the relative jurisprudence and of administrative devices useful to the inhabitants both to participate in decisions on urban regeneration and to contribute from the bottom up with shared care actions. This chapter straddles the boundary between urban planning and sociology instead, working on the common ground of rights: the traditional right to public services on the one hand and the more recent right to care for the commons on the other. The attempt here consists in the critical analysis of one of the fundamental public services: the school. We start from a territory that we have studied, analyzed, and practiced both as scholars and as practitioners.

11.2 The Case Study of Settimo Torinese

11.2.1 Schools and Proximity Areas. Creating Settlement Conditions for an Innovative School-City-Territory Project

Settimo T.se is a municipality of the metropolitan city of Turin located in the capital's first belt, whose territory is divided almost equally between anthropic (49%) and agricultural (47%) land uses and land cover. In comparison, the presence of natural systems (2%) and water bodies (2%) is minimal (LCP 2021).

The settlement system has developed according to a radio-centric expansion model with few widespread features and a robust system of mobility infrastructures of varying functional rank. To sum up, it can be traced back to two main macro-environments: the compact central urban area with concentric development (17% of the municipal territory) and the suburban territory, arranged almost as a crown around the concentric area.

The endowment of existing planning standards in Settimo (2018) covers about 10% of the municipal area (3.1 million sq.m.), distributed as follows (Table 11.1): almost 60% for social services at the municipal level (Art. 21, Law 56/1977) and the remainder for general interest services (Art. 22, Law 56). These endowments guarantee, concerning the resident population in 2021 (47,006 inhabitants), far more than the minimum quantity provided by the Piedmont spatial planning law.

School facilities below compulsory education constitute an infrastructure with a high capillarity and diffusion, covering about 11% of the total surface area at municipal level standards, to which two higher-level schools are added. This condition enhances their possible role in rethinking the spatial relations between school and city, with a view to openness, inclusiveness, and multifunctionality of the school infrastructure. Recognizing that school takes on a civic role by continuously interacting with its social and territorial context (Ministry of Education 2022), the spatialization of the school heritage and the knowledge of the conditions of the settlement context in which it is inserted are indispensable for recognizing its actual potential.

Table 11.1 Existing planning standards in Settimo T.se

Existing planning standards (Law 56/77)	Surface area (m^2)	Incidence on municipal territory (%)	Endowment per inhabitant (sq.m./inhab.)
Municipal level services (art. 21)	1.803.166	59	38
General interest services (art. 22)	1.278.486	41	27
Total (artt. 21 and 22)	3.081.653	100	65

Source elaboration of V. Vitulano based on Città di Settimo T.se (2018)

The construction of a census of planning standard endowments (of which schools are part) constitutes the first fundamental step in undertaking decision-making processes based on reading and interpreting the site-specific characteristics of the territory and the community. Furthermore, as the presence and distribution of school differ, it is possible to assume that the radii of action and the possible influences they can generate are equally different. For these reasons, the concept of proximity has been assumed as a spatial paradigm with which to address topical urban, social, and civic issues such as rethinking the virtuous interaction 'school-city' and the relations between different types of urban public spaces, caring for the wellbeing and health of citizens and protecting the urban environment through the enhancement of public green areas in school lots and their proximity.

Therefore, 'proximity areas' have been operationally defined through the generation of a buffer zone of 300 linear meters from the boundaries of the school lot; this distance identifies a walking distance of approximately 5 min and is considered the ideal distance to allow independent mobility for children between 6 and 12 years of age (UNICEF 2018; Barò et al. 2021).

The definition of school proximity areas is functional to the analysis of the characteristics of this urban area, i.e., the conditions of the space outside the school fences, especially in relation to the presence or absence of other public services. For example, data on the incidence of existing planning standard areas within the proximity areas show that, out of a total of 27 school buildings, in about 3/4 of the situations, the incidence of other services is less than 21%; the rest of the cases show values between 21 and 30%. Among the latter, the schools in the Borgo Nuovo District stand out, confirming the presence of more significant endowments in the parts of the territory that are the result of the most recent growth processes; on the contrary, most of the lowest incidence values refer to the schools located in the portion of municipal territory between the Turin–Milan railway and provincial road n. 11.

Since the 0–6 age group is of particular importance for public education policies, let us focus specifically on the proximity areas of kindergartens (Fig. 11.1). In this case, the situation is in line with what was stated before most of them remain under 21% of areas for services of both municipal and general interest.

Green public services are particularly scarce in most proximity areas (24 cases out of 27), where the incidence is less than 8%; this evidence is also confirmed for kindergartens. The most 'virtuous' cases are, again, located in Borgo Nuovo, where, moreover, the administration is concentrating its planning efforts to launch a qualified transformation in terms of functions and new opportunities for its inhabitants, completing the regeneration process begun in the late 1990s (Città di Settimo T.se 2018).

It is, therefore, possible to note that in Settimo the strong point of a network of public services around schools in most cases is not given by the presence of green standards—especially green with high ecological qualities, given the high presence of sports facilities, this due to the typical morphology of a monocentric municipality strongly urbanized in the central part compared to the rest of the territory (Fig. 11.1).

The data also reveal a series of typical situations where the low incidence of permeable green spaces occurs, especially near those polarities of services 'built'

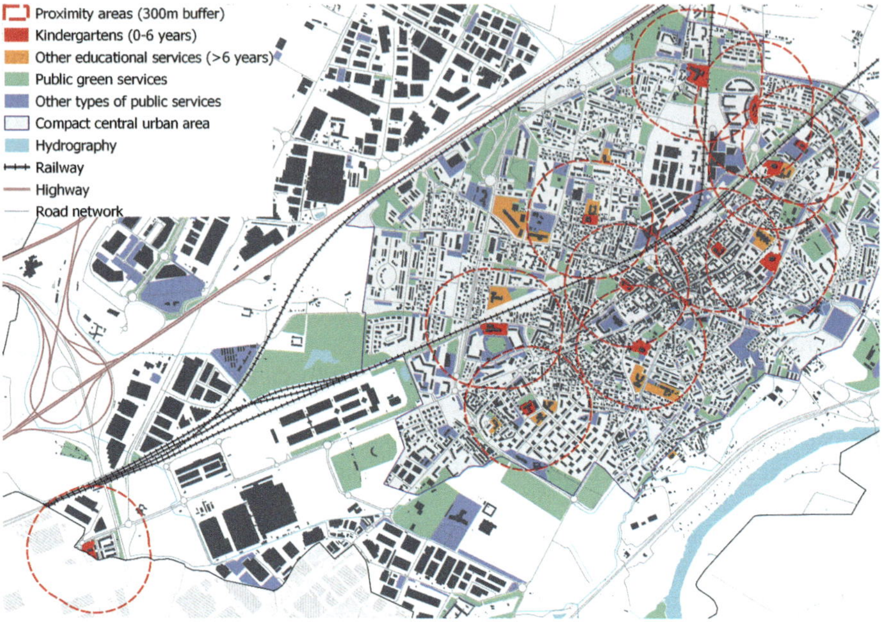

Fig. 11.1 Proximity areas of kindergartens in Settimo Torinese. *Source* elaboration of V. Vitulano based on Città di Settimo T.se (2018)

within the concentric area, where clusters of schools and other services (such as car parks and facilities of common interest) have been created. Here the medium–high incidences are mainly explained by a concentration of schools (of different grades) that could give rise to the hypothesis of establishing specific school zones within the meaning of the Italian Road Code.

In many cases, public interventions on school buildings that have been put in place seem to be mainly oriented toward safety and efficiency enhancement, also using funds from the National Recovery and Resilience Plan (PNRR). However, in order to compensate for the low presence of public green areas within a now densely urbanized fabric, where it is difficult and costly to envisage the creation of new porosity, an element on which to pay more attention may be precisely the adjacent lot of school buildings, verifying the presence of adequate open spaces with quality characteristics, also in terms of existing vegetation and trees.

11.2.2 An Established Participatory Tradition and a Future for the Community in Action

From the point of view of participatory urban planning, the context of Settimo T.se is prototypical of a style of government that spanned the 1990s and 2000s in the name of political-party continuity based on pluralistic and inclusive urban regeneration process. These were conceived by the local public administrators, also driven by similar supra-local policies such as regional, national, and European ones for the regeneration of the city and the territory (Ciaffi and Mela 2011). In the Turin metropolitan area, there are also other municipalities known and studied for this participatory approach, which have expressed different social vocations from time to time, for example centered on participatory youth policies or on the co-management of green areas based on the informal groups of older people or, again, on the involvement of groups of parents in school policies as protagonists. When the local government style is based on a consolidated participatory matrix, social innovation based on the new right to care for common goods faces a challenge on several fronts. The main one concerns the transition from a bipolar paradigm to a collaborative one: from the administrator-administered relationship, which can also be characterized by a cooperative spirit but the helm is always substantially held by the elected politicians, to an active community that includes both administrators and those administered. In the latter subsidiary perspective, not only are decisions taken through horizontal governance but urban regeneration is done by co-managing common goods together. These are processes that basically arise from the autonomous initiative of individuals who want to contribute by carrying out actions of general interest, and want to be favored in this purpose by the public administration: This dynamic is very different from accepting an invitation to participate in an urban regeneration process designed by public officials.

Attracted by this new horizontal style of government, the municipality of Settimo T.se adopted the Regulation for the shared administration of common goods in 2021, dedicating a section of its website and identifying human resources and a reference office. However, the signing of collaboration agreements requires a certain amount of time to mature. The recent pandemic crisis certainly does not help, but we would like to reflect here on a historical turning point in progress, which consists in the transition from services traditionally intended to 'hybrid and shared services' (Ciaffi 2020). There are many points of contact with the recent debate on proximity services, starting with not believing in a future society that can give up public spaces and services in the name of 'completely to/from home': a feasible undertaking on the one hand by designing the conditions for a suitable environment from an urban and social point of view, on the other hand by producing events and initiatives that lead to the generation of new communities (Manzini 2021).

As part of the research group of the project 'The city goes to school. School squares as spaces of quality and environmental sociability' (Pileri et al. 2022), we investigated the relational and social dimension of the school as a common good, or as a product of the interaction. School has been considered as an institution and public service, but also as a common space of a multiplicity of territorial subjects that in various capacities participate in the educational experience of minors, contributing skills, resources, and energies. This process co-produce, in fact, training opportunities and educational spaces, complementary to the traditional ones. Listening through in-depth interviews allows us to observe, in the perspective of horizontal subsidiarity, some dynamics underway also in Settimo T.se. We leave the floor to three qualified witnesses because we believe that their voice well describes the process of hybridization underway in the educational sphere together with the need to share responsibilities between public subjects such as the head teacher of a local comprehensive school, private subjects such as the president of a community foundation, and third sector subjects such as the president of the oratorie's social promotion association:

> In the meantime, I think that the educating community is something that has really begun to be built in recent years. But this in general, not just in Settimo, because the concept of school has changed a bit. Thinking only of school management, first there was the apical role of the principal, who was far and away, who organized everything and there was the school, there were the teachers, grades and stop. In primary school [...] the children, spending more time at school, also did things, activities [...] opening up to experience in the area. But this did not happen in schools, for example middle and lower secondary schools. And instead, slowly over the years, where there are slightly more enlightened schools, and certainly also starting from the experiences of southern Italy, of schools that really tried to prevent juvenile delinquency in difficult neighborhoods, an attempt was made to start creating the school as a place of reference, and consequently with reference adults, who may not necessarily be just the teachers [...] so we really started to think about what an educating community is and what does it mean. Because the border lies precisely in the inside of the school and outside the school: it is on that border point that the game is played.
>
> (Maria Zindato, Headmaster I.C. Settimo T.se).
>
> In Settimo, we have enough squares: some historic, others created during urban redevelopment in the city [...] In this period the squares, even during the lockdown period, were populated by kids who, as soon as they could, reappropriated them in a more or less, shall we say, correct. Here, if we read it as a problem there, we are lost. [...] If we read it from an educational perspective, it becomes a potential. Difficult, very hard, whatever we want, but it's the only perspective, as far as I'm concerned, that's it.
>
> (Tiziana Tiziano, President of the Solidarity Community Foundation, Settimo Torinese).
>
> It's not easy, but all in all we have more than 250 members, we already have about sixty educators in the field, more responsible and adults who are helping out. So, I would say that the speakers are still there in the city of Settimo and they are still active. Then we have some initiatives in addition to what is the classical oratory [...] I believe that in all of Italy the oratory is not absolutely dead [...] Because if we reduce the oratory to just the opening of the courtyard, that's a long time; if you experience the oratory as a training ground for life, which looks on many fronts, it is certainly an enriching experience.
>
> (Luca Spiga, President of the Aps Oratories of Settimo Torinese).

11.3 Conclusion

There is a strong awareness that the project of a new urbanity, capable of interpreting a more performance-based declination of urban quality, cannot be satisfied within the perimeter of some quantitatively measurable parameter (to which, moreover, we recognize a certain precondition), and that it is necessary to return to giving urban spaces a more significant and widespread porosity, permeability and accessibility. Furthermore, it is necessary to promote an updating of the cultural and disciplinary sense of the public (and common) heritage constituted by standards, indicating possible adjustments of specific contents and directions for the new needs and demands expressed by cities, communities, and activities.

Acting on the heritage of public endowments requires parallel work on three fronts of action:

- the definition of needs and their localization and spatialization;
- the design and implementation of spaces and equipment to meet needs;
- the management of spaces and equipment.

This leads to several operational proposals. In particular, the realization, as a compulsory content of urban planning to guarantee the basic levels of services, reconnaissance, and verification, through public participation processes, of the state of affairs regarding urban planning standards (levels of services offered and expected performance objectives). A participation process aimed at contributing to the better quality and effective localization of public spaces, the correct sizing of urban allocations, the enhancement of the potential present in the reference territory, and the establishment of reference and governance frameworks for urban regeneration processes.

So why talk about hybrid and shared services? On the one hand, sharing is motivated by the fact that it is now even more evident to us that many services, which more and more people perceive as common goods of general interest, have found themselves facing challenges that we cannot continue to think are faced only by public managers. It is the society of care, which not only uses and consumes spaces and services, but which wants to deal with them beyond the twentieth-century model, but in the conviction of a State that must not retreat, on the contrary, is increasingly called to be a true ally. On the other hand, we talked about hybridization because the desire to contribute together to solve problems, each with their own know-how, is growing when it is possible to channel it into legal devices, such as pacts and alliances between public managers with technical and political roles, private subjects, associations, and informal groups, but also single individuals.

Obviously, there are many cross-disciplinary issues that remain open. For example, to remain in the school field, how to define the so-called educating community? What are the actors that compose it, the roles, skills, and relationships that bind them and how do they use and modify the urban space? How is the educational service reconfigured and what characteristics does it assume, including spatial ones, if one reads the collaborative practices of construction of the educational offer? What is the perimeter of the school, in terms of use of space and construction of the service, if one observes the action of the educating communities in the survey areas?

Acknowledgements The text is the result of multidisciplinary work coordinated and shared between the authors in which paragraphs 1.1 and 2.1 are attributed to Carolina Giaimo and Valeria Vitulano; paragraphs 1.2 and 2.2 to Daniela Ciaffi and Emanuela Saporito; and paragraph 3 to Carolina Giaimo and Daniela Ciaffi.

References

Arena G, Bombardelli M (ed) (2022) Amministrazione condivisa, Editoriale scientifica italiana, Naples

Baró F, Camacho DA, Pérez Del Pulgar C, Triguero-Mas M, Anguelovski I (2021) School greening: right or privilege? Examining urban nature within and around primary schools through an equity lens. Landsc Urban Plan 208:104019

Ciaffi D (2019) Diritto a prendersi cura dei beni comuni: il massimo della partecipazione. In: Labsus. 16 Sept 2019

Ciaffi D (2020) Servizi ibridi e condivisi, per prendersi cura dei beni comuni. in Labsus. 9 Jan 2020

Ciaffi D, Mela A (2011) Urbanistica partecipata. Dimensioni, spazi, strumenti, Carocci, Rome

Ciaffi D, Saporito E (2022) Il diritto alla cura dei beni comuni come palestra di democrazia. Sociologia urbana e rurale n.1, pp 39–51

Città di Settimo T.se (2018) Variante di revisione generale, Proposta tecnica di progetto preliminare, Relazione tecnica

Giaimo C (2021) (Ri) comporre relazioni. Le dotazioni urbanistiche. In: Moccia FD e M. Sepe (eds) Benessere e salute delle città contemporanee, INU Edizioni, Rome, pp 271–278

Giaimo C (eds) (2022) A scuola di spazio pubblico. Visioni, strumenti, azioni per un nuovo rapporto scuola—città a Torino, INU Edizioni, Rome

Labsus (2021) Rapporti sull'amministrazione condivisa dei beni comuni 2017, 2019, 2021 available on labsus.org

Lefebvre H (1970) Il diritto alla città, Marsilio, Venice

Mazza L (2015) Spazio e cittadinanza. Politica e governo del territorio, Donzelli editore, Rome

Manzini E (2021) Abitare la prossimità. Idee per la città dei 15 minuti, Egea, Milan

Ministero dell'istruzione (2022) Futura. Progettare, costruire e abitare la scuola.https://pnrr.istruzione.it/wp-content/uploads/2022/05/LineeGuida_ScuolaFutura.pdf

Pileri P, Renzoni C, Savoldi P (2022) School squares. Reinventing the dialogue between school and city, Corraini, Milan

UNICEF (2018) Shaping urbanization for children a handbook on child-responsive urban planning. https://www.unicef.org/reports/shaping-urbanization-children

Part IV
Digital Tools

Chapter 12
The 3D Metric Survey for the Digital Cartographic Production to Support the Knowledge of the New Municipality of Mappano

Egle Beani, Elisabetta Colucci, Luigi La Riccia, Andrea Maria Lingua, Paolo Felice Maschio, Francesca Matrone, Alberto Possa, and Ammj Traore

Abstract The complexity of the territories and cities is the genesis of the need to carry out spatial analyses, simulations of complex phenomena and urban planning, requiring an evolution of very large-scale numerical cartography. The representation of cities has not to ignore some essential aspects as well as: the three-dimensionality of the land and all its natural and artificial objects, the multi-scale of geometric and descriptive information to allow different levels of detail and granularity according to the operational purposes, and the interoperability, which enable sharing of map data among various stakeholders and applications for complex analysis. The outcome

E. Beani · E. Colucci (✉) · A. M. Lingua · P. F. Maschio
DIATI—Department of Environment, Land and Infrastructure Engineering Politecnico di Torino, Turin, Italy
e-mail: elisabetta.colucci@polito.it

E. Beani
e-mail: egle.beani@studenti.polito.it

A. M. Lingua
e-mail: andrea.lingua@polito.it

P. F. Maschio
e-mail: paolo.maschio@polito.it

L. La Riccia · A. Possa · A. Traore
DIST—Interuniversity Department of Regional and Urban Studies and Planning Politecnico di Torino, Turin, Italy
e-mail: luigi.lariccia@polito.it

A. Possa
e-mail: alberto.possa@studenti.polito.it

A. Traore
e-mail: ammj.traore@polito.it

F. Matrone
DISEG, Department of Structural, Geotechnical and Building Engineering, Turin, Italy
e-mail: francesca.matrone@polito.it

G. Brunetta et al. (eds.), *Post Un-Lock*, The Urban Book Series,
https://doi.org/10.1007/978-3-031-33894-6_12

resulted in the evolution of traditional cartography into numerical cartography toward new instruments and applications such as digital twins. They constitute a virtual replica of physical, potential, and actual resources equivalent to objects, processes, people, places, infrastructures, systems, and devices related to the city and the territory. The case study selected for this research is the recent municipality of Mappano, in the metropolitan city of Turin. The methodology consists of a semiautomatic extraction of cartographic information and geometry essential for defining a digital twin. It has been developed investigating the existing Piedmontese regional geoportal (BDTRE) datasets and data models, and the Geographic Information System (GIS) standard specification for 3D city models, CityGML, mainly considering the first three levels of detail (LoD). The methodology and the resulting spatial data application have been validated by in-loco surveys and digital tools in GIS environment. The starting base map entities derived from a high-resolution acquisition from drone photogrammetry (UAV, Unmanned Aerial Vehicle, platforms have been adopted). The outputs of the aerial survey, orthophotos, and dense altimetric models (DTM, Digital Terrain Model and DSM, Digital Surface Model) were essential to extract various cartographic information.

Keywords 3D metric survey · GIS · Geo-topographic database · Digital cartography · 2D/3D city models

12.1 Introduction

Smarting the city is an intricate process due to the complexity of a city. The city is not an automated system that can be easily understood and predicted, but rather a living system that evolves every day through variations and developments of its physical constructs, economic and political activities, social and cultural settings, and ecological systems (Shahat et al. 2021). The knowledge of a territory is strongly related with the accessibility to a complete and detailed digital cartography in which quantity and quality information can be collected and organized. Data harmonization must play a key role to guarantee interoperability of these data, and hence, policies have been advanced at European and global levels to pursue this aim. Data standardization is part of the process for better planning and policy evaluation, more efficiency, optimization and innovation. Furthermore, the creation of a European data infrastructure can simplify the sharing of spatial information between public administrations, facilitate public access to environmental spatial information throughout Europe, and assist in decision-making processes relating to the environment and the territory (Geoportale Nazionale 2021).

The infrastructure for geographic information (GI) is the set of spatial data, metadata, network services and technologies, policies and institutional agreements aimed at sharing, accessing, and using to facilitate the availability and homogeneity of geospatial data. Legislative Decree of January 27th, No. 32/201022 for "Implementation of Directive 2007/2/EC, establishing an Infrastructure for Spatial Information in the European Community (INSPIRE)"[1] was published by Italy aimed at the creation of a national infrastructure for spatial information and environmental monitoring allowing the Country to participate at INSPIRE. Services are available to the public free of charge, but the State might limit (art. 13) public access to spatial datasets and services for many reasons such as public security or national defense.

Moreover, also at the international level, the GIS standard data model for 3D city representation is CityGML. It extends the GML, by adding the semantic values for the 3D representation of city objects (Gröger et al. 2012). The standard implements five levels of detail (LoD) for a multi-scale model of spatial entities representation. LoD 0, deals 2D or 2.5D objects (scale about 1:50,000—1:10,000). LoD 1 represents buildings in 2.5D or 3D (1:25,000—1:10,000). LoD2 and LoD3 represent city district or architectural models (1:10,000—1:5000 and 1:2500, 1:2000 in Italy—1:1000). Finally, LoD 4 is dedicated to the interior building (1:1000 or 1:500) (Biljecki et al. 2016).

In the national framework, each Italian Region must own spatial datasets, related to the specific geographic area of pertinence, collected in a geoportal providing public access to these services organized in standardized spatial information of the territory. Piedmont region has BDTRE, initials of spatial database reference for the authorities—Base Dati Territoriale di Riferimento degli Enti—which is the geographical database of whole Piedmont promoted by the Piedmont region authority, whose contents are structured according to the national "Technical rules for the definition of the content specifications of the geo-topographic databases" (MD 2011/11/10) primarily aimed at supporting the planning, governance, and protection of the territory. The BDTRE therefore assumes the role of shared and free-to-use storage of all spatial data, from which Regional Technical Cartography (CTR) derives. The region has published a standard's specification document, to produce topo-cartographic database pursuant to the MD 2011/11/10 and to support cartographic productions, which constitutes a base for public administrations on spatial information, according to Art. No.59 of Legislative Decree. No. 82/2005 (Digital Administration Code[2]).

All municipalities must guarantee the development of a database (DB) containing spatial and semantic information of their objects. This set of data help also the different stakeholders of the city involved in many activities such as urban planning, policy-makers actions, restoration, and revitalization of marginal areas. Therefore, the tasks of the newly formed municipalities include the creation of a geodatabase containing all the related information. During the DB design, all the three-dimensionality objects in the land and all its natural and artificial objects have to be considered. Moreover, the datasets have to be structured with a multi-scale, multi-level, and multi-granularity approach. The result is a numerical cartography designed with innovative methods, instruments, and applications such as digital twins, a digital representation of the reality, in 2D or 3D.

[1] "Implementation of Directive 2007/2/EC, which establishes an infrastructure for territorial information in the European Community (INSPIRE)".

[2] Code of digital administration.

The development of accurate and precise large-scale numerical 2D cartography integrated with the production of 3D city models is increasingly studied worldwide and required by public administrations. Therefore, the municipalities have a digital model of the territory, also in 3D, of significant potential.

The methodology here presented consists of a semiautomatic extraction of cartographic information and geometry essential for defining a digital twin. The cartographic production derived from a rigorous 3D metric survey after an accurate analysis of the Piedmont regional geoportal (the BDTRE, Base Dati Territoriale di Riferimento degli Enti piemontesi[3]) datasets and data models, and the GIS standard specification for 3D city models, CityGML. The case study selected for this research is the recent municipality of Mappano, in the metropolitan city of Turin. The methodology and the resulting spatial data application have been validated by in-loco survey and digital tools in GIS environment. The starting base map entities derived from a high-resolution acquisition from drone photogrammetry (UAV, Unmanned Aerial Vehicle, platforms have been adopted). The outputs of the aerial survey, orthophotos, and dense altimetric models (DTM, Digital Terrain Model and DSM, Digital Surface Model) were essential to extract various cartographic information.

12.2 The Case Study

Mappano is a town in the first belt of the metropolitan city of Turin not far, less than five kilometers, from the Turin Stura railway station. It is characterized by being defined as the youngest municipality in Italy as it was officially formed in 2017. Mappano was born from "the supremacy of the local identity [which] is the main driver of social actions. [...] Until the time of the referendum in 2012 there was no official, unique, or shared definition of what the territory of Mappano is, a conurbation that extends over the territory of the municipalities of Borgaro Torinese, Caselle Torinese, Leinì, Settimo Torinese, and Torino" (Pileri and Scalenghe 2016). The actual boundaries of the city were established after the referendum on November 11, 2012 (Ibid.). This "administrative fragmentation," occurred with the birth of a small new municipality, brought environmental issues up, since at greater number of units governing a territory, each of them maximizing its own goal on a small territory, increase the need for inter-agency dialogue and coordination (Ibid.) other than difficulties in finding a shared line of planning view, time, and costs.

As above mentioned, due to the necessity to organize and spread a digital cartography of the municipality, Mappano has signed a collaboration agreement with the DIATI and DIST departments with the aim of creating a highly accurate, updated, and complete cartography, essential for the implementation of its first General Town Plan (Italian *Piano Regolatore Generale*). At the basis of this commitment, there is the recent born of Mappano and therefore the consequent assignment of adopting a General Town Plan, as provided by law no. 1150 of 1942. Correcting the flaws of the

[3] https://www.geoportale.piemonte.it/cms/bdtre/bdtre-2.

forced coexistence of the plans of four municipalities (Caselle Torinese, Borgaro, Leinì e Settimo Torinese) has been the starting point of the plan. Hence, the differences of quality among the composed Mappanese cartography have increased the interest in harmonizing the existing data and acquiring a new one. The methodology here presented describes this process.

12.3 Methodology

As shown in Fig. 12.1, the workflow of the methodology is divided into three main phases: the data acquisition, the processing, and the definitions of outputs. The first step consisted of a 3D metric survey adopting traditional topographic methods (GPS/GNSS) and photogrammetry techniques. The first step was measuring the topographic network to obtain the vertices coordinates. Then, UAV aerial photogrammetry has been adopted considering different details and scale representation levels. For this reason, two different drones have been adopted: the fixed-wing drone eBee Plus by senseFly for the territorial acquisition and the DJI Phantom 4 RTK for the city objects survey. After the acquisition phase, the data have been processed following consolidated approaches to generate 3D models in point clouds, meshes, digital models, and orthophotos. The data outputs, as described below, are all the outcomes of the post-processing phase: the spatial objects (both vectors and rasters, such as digital terrain models) included in the new digital twin from the BDTRE Geo-topographic database of the Piedmont region.

12.3.1 Data Acquisition

The initial inspection of the area allowed to proceed with the preliminary phases of the photogrammetric survey. First of all, markers on the ground that make up the topographic network were placed uniformly on the area to be surveyed. These markers, also named ground control points (GCPs), are essential for returning the final data, allowing the correct scaling and georeferencing of the outputs. In this case, 60 $\times$ 60 cm panels in semi-rigid plastic or high-strength PVC fabric were used, characterized by high-contrast colors that allow them to be easily recognized from the UAV images (Fig. 12.2a). Where possible, natural points have been used, such as manholes, manhole covers, or fixed objects that are clearly identifiable and permanently present in the surveyed area (Fig. 12.2b).

Figure 12.2c shows the final distribution of control points materialized in the field, in particular 30 GCPs evenly distributed over the area. The position of those GCPs was acquired using a GNSS receiver (Fig. 12.2c). In particular, ItalPOS SmartNet GNSS network was used: a service owned by Leyca Geosystems which offer the correction of measurements for positioning in RTK with fixed bases throughout the national territory.

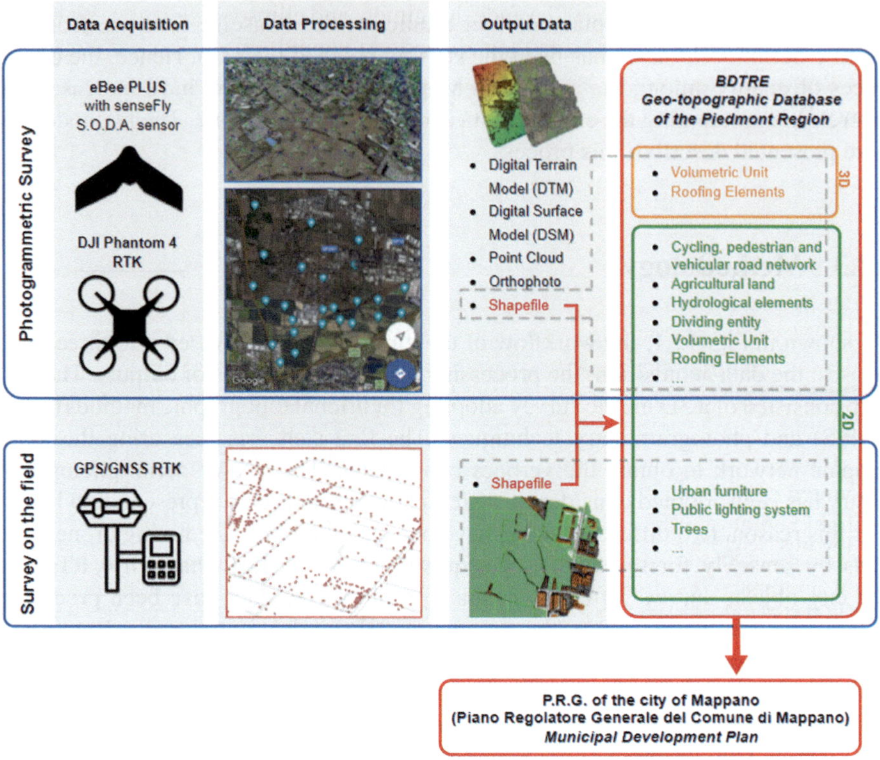

Fig. 12.1 Methodology used for the creation of the new municipal cartographic products. The output data obtained from the photogrammetric and field survey merged into the Geo-topographic database and served as the basis for the definition of the new Municipal Development Plan

Fig. 12.2 GCPs materialized in the surveyed area and GNSS receiver (**a**). Natural point selected as GCP (**b**). Overall view of the GCPs (**c**). (Images from the 3D metric survey carried out by the Geomatics group of Politecnico di Torino)

The coordinates in terms of latitude, longitude, and ellipsoid elevation (with reference to the ellipsoid WGS4 and then converted to geoid) of all the GCPs were thus acquired.

Photogrammetric survey

A strategy of multiple drone flights was implemented to cover the entire area of 11.896 km^2.

The system used consists of a fixed-wing aircraft called eBee Plus produced by senseFly. The aircraft weighs about 1.1 kg including camera and battery, with a wingspan of 110 cm. The cruising speed varies between 40 and 110 km/h, with a wind resistance of up to 45 km/h. The flights have been planned around 40 min, in order to avoid running out of batteries before the scheduled landing.

In this application, the camera supported by the senseFly SODA device was used with 13.8 $\times$ 8.8 mm format RGB sensor, 10.6 mm focal length, for a total of 20 megapixels, global shutter, and equipped with dust and shock protection. The mounted camera allows for a theoretical ground resolution of about 2.9 cm/pixel for a flight height of 120 m.

The complete coverage of the municipality of Mappano was achieved through a total number of 13 flights (Fig. 12.3). A total of 5.646 frames were acquired, at a flight height of about 140 m, with a resolution of 3.16 cm/pixel.

During the acquisition phases, the images have been taken from different angles: nadiral and oblique. The latter are used to document the areas not visible from the former (as façades) and to develop the three-dimensional model with a higher level of detail.

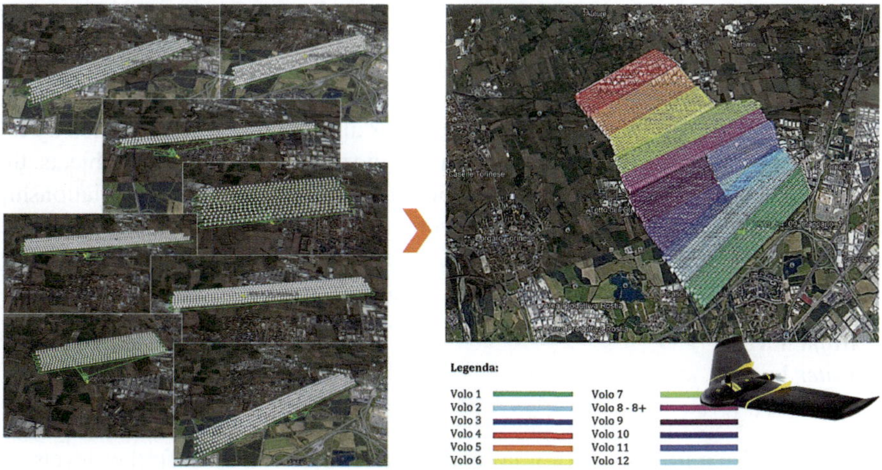

Fig. 12.3 Examples of flights performed for the survey (left) and their union (right). On the right bottom, the eBee Plus UAV used. (Screenshot from the software used for the planning of the UAV survey)

Fig. 12.4 On-site survey in Municipal of Mappano took place in January 2021 and it involved the direct team, scholarship holders, and PhD students from DIATI. (Images from the 3D metric survey carried out by the Geomatics group of Politecnico di Torino)

Survey on the field

An on-site survey was necessary to acquire all those elements not detectable from the orthophoto analysis, such as street/urban furniture (MN_ARR), vertical road signs, single trees, and artifacts such as traffic lights, control groups (MN_INT), electrical network/gas cabins (MN_IND) (Fig. 12.4). To this aim, Leica GS14 and GS18 GNSS RTK rover were used, fixing the position when under 4/5 cm of accuracy.

12.3.2 The Geo-Topographic Database

As anticipated, the Data Catalog (Simplified Operational Specifications for Geo-topographic DBs for use by public administrations—Piedmont Region, Version 2.2.1, April 5, 2017) can be considered as a real translation of the structure illustrated in the CityGML model, since the reference structure is also in this case constituted by the *class*, which defines the representation of a specific typology of territorial objects: the properties, the data structure, the acquisition and structuring rules and the relationship with the other objects. Within this framework, four levels are identified to define the topological land cover (Fig. 12.5):

urban level,
traffic level,
water level,
vegetation level.

There is also a fifth level for non-topological classes and three further levels for graphs and addresses: viability graph level, hydrography graph level, and address level.

To distinguish the type of attributes, identification codes have been used within the specifications, determining the creation of an external *dictionary*.

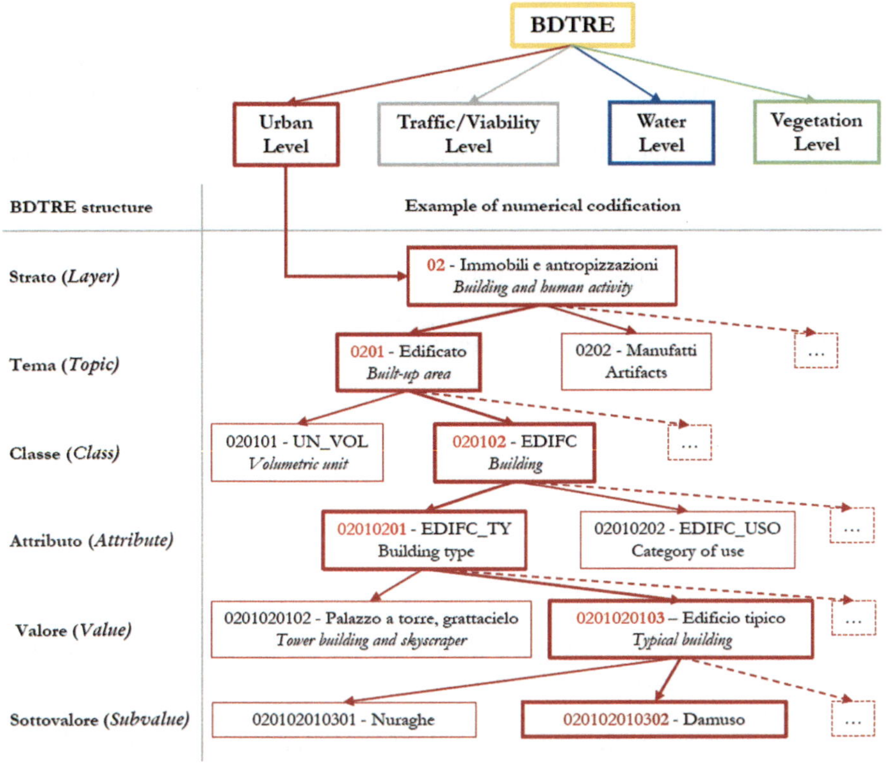

Fig. 12.5 BDTRE structure followed for the data output. (Schema designed by the authors)

Urban level

Within the urban level, there are multiple classes such *as equipped areas, buildings, minor buildings, industrial artifacts, architectural details, street furniture,* and so on. Specifically, as regards the Building Class (EDIFC—020,102), the related attributes that have been inserted are:

the building type (02,010,201 EDIFC_TY): generic '01', tower building '02', typical building '03', etc.;

the category of use (02,010,202 EDIFC_USE): residential '01', administrative '02, public service' 03 ', military' 04 ', etc.;

status (02,010,204 EDIFIC_STAT): under construction '01', ruined '02', or built '03'.

Traffic/viability level

Within the traffic/viability level, different classes are identified corresponding to the different types of traffic areas: vehicular (AC_VEI), cycle path (AC_CIC), pedestrian (AC_PED), mixed (AR_VMS), and so on.

Water level

The Water level of the BDTRE database is defined as the classes that provide a representation of the hydrographic system of the Piedmont region. This level has been identified different elements such as river/stream or canal (EL_IDR), the different types of standing water (SP_ACQ) and the water node (ND_IDR), i.e., the point of intersection between water elements.

Vegetation level

In the vegetation level, the classes relating to the different types of vegetation present are identified: green areas (AR_VRD), agricultural crops (CL_AGR_TY), forest (BOSCO_TY) and pastures and uncultivated (PS_INC_TY).

12.4 Results

12.4.1 Cartographic Products

As regards the processing of photogrammetric data, it took place at the Geomatics Laboratory of the Department of Environmental, Territory and Infrastructure Engineering of the Polytechnic of Turin. Agisoft Metashape and Pix4D were chosen as processing software, allowing a data comparison (Beani 2019). The procedure followed is rather automatic, since it consists of sequential steps (generation of sparse point cloud, dense point cloud, DEM, orthomosaic and eventual 3D textured mesh); however, it requires a careful check on the residual errors of the 24 GCPs used. This workflow provided mean errors of 2.15 cm with accuracies of around 1.52 cm and 1.4 cm for the horizontal coordinates (x and y) and 0.6 cm for the vertical coordinate (z), allowing to generate large-scale cartographic products up to 1:1000.

The above-described pipeline outputted the required cartographic products in terms of dense cloud (12.424.864 million points), orthophotos (Fig. 6a), and digital surface model (DSM) (Fig. 6b) with a resolution of 4 cm/pixel. The processing of this data required a computational time of about 5 h to obtain the dense cloud in *average* quality, 18 h for the orthophoto, and 33 min for the DSM. Once obtained the DSM, a digital terrain model (DTM) was created through the automatic point cloud classification of the Metashape software. This latter tool allowed to speed up the exclusion of categories such as buildings, vegetation, or vehicles, in order to keep only ground points; however, a manual intervention was anyway required.

The DSM (coordinate system WGS 84 / UTM zone 32N, EPSG: 32,632) has a color legend from red to blue that identifies the altitude of the points of the territory from the highest in the north-west to the lowest in the south-east, in correspondence with the area of the hydraulic work known by the community as the "spillway."

In general, an average GSD of 316 cm was obtained for the whole area.

In addition to these products, data have been processed from on-site survey. Hence, street furniture was classified following BDTRE standards in one data (MN_ARR).

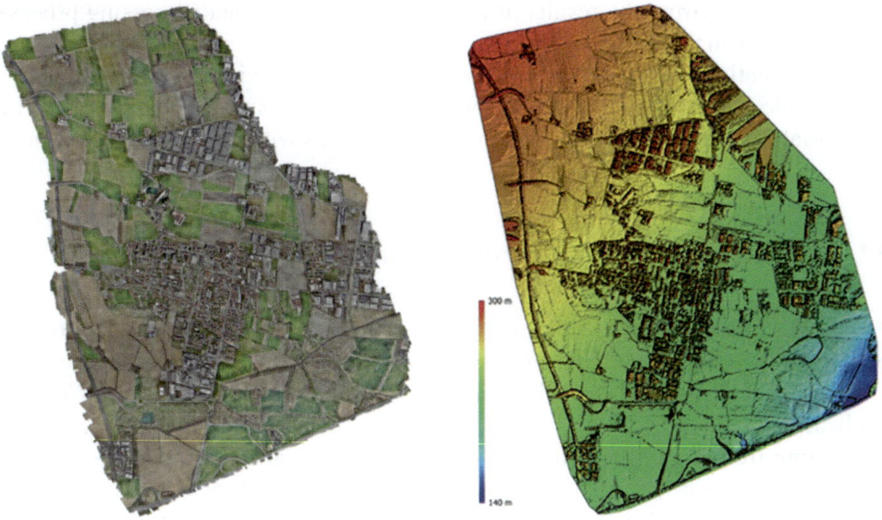

(a) (b)

Fig. 12.6 Orthophoto (**a**) and corresponding DSM (**b**). (Data products from the post-processing carried out by the Geomatics group of Politecnico di Torino)

Figure 12.7 shows the result of this process. This information can be used for a plurality of tasks such as updating the level of maintenance of each artifact and their extension on the territory.

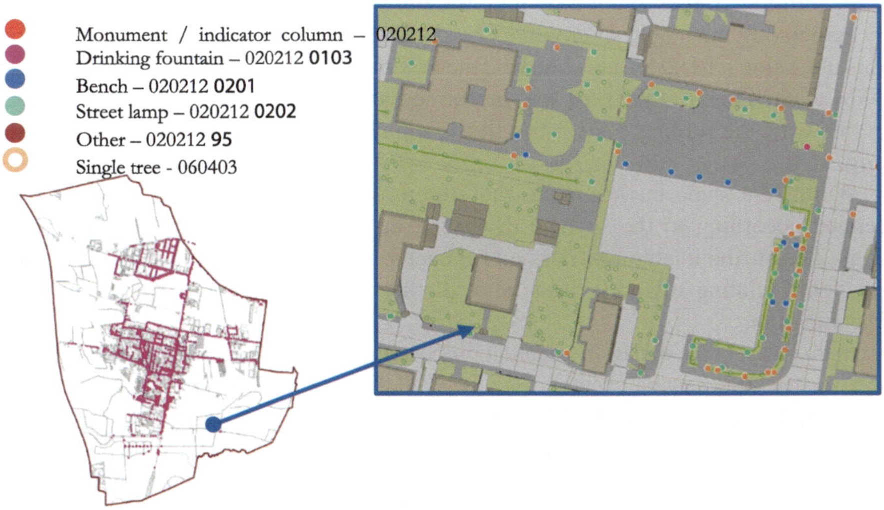

Fig. 12.7 Focus on Piazza Don Amerano showing street furniture's classification result

Then, starting from the results of the photogrammetric data processing process and the survey on the field, in the general context of 3D data generation it is necessary to focus attention on the objective of the project, which envisages obtaining a geo-topographic database based on the Simplified Specification standards proposal for the Territorial Reference Database of Management Bodies within the Piedmont region.

12.4.2 The Digitalization Process and the BDTRE

As regards the geometric definition of the database entities, the outputs of the photogrammetric process and the survey in the field constituted the basis for its digitalization. This operation took place in the work environment of the ESRI ArcMap application.

Starting from the orthophoto, used as a reference image, 2D modeling consists of the generation of geospatial vector data, by defining polygons associated with different layers, which represent the elements relating to the classes of interest. These files, called shapefiles, were the second output of the project.

This is a rather simple manual procedure, but relatively time-consuming. The level of geometric detail for most of the shapefiles stopped at the definition of the contours, therefore a two-dimensional LOD0 (Fig. 12.8). For many of these layers, in the phases following the digital restitution, further inspections were carried out aimed at improving the level of detail, e.g., the insertion of information relating to the balconies (included in the architectural details shapefile—PAR_AR—and not obtainable from the point clouds) or the projections of the roofs, which made it possible to more accurately derive the footprint of the buildings on the ground.

In addition, a 3D shapefile (volumetric units) was also created for the buildings by associating the height interpolated by the DSM with a centroid for each volume, thus obtaining a LOD1. Subsequently, the association of a height interpolated by the DSM to each pitch of the roof was performed and permitted the creation of LOD2. In this way, the roofing system for all the municipality was directly defined in 3D (Fig. 12.8).

Within the same environment, the structure of the desired database was also defined according to the scheme described in paragraph 2.2 and related subsections. The attribute table was defined for each polygonal element modeled, creating the fields relating to level, class, and all the attributes required by the BDTRE Specification.

In this way, at the end of the manual processing in ArcGIS-ArcMap, an exportable and further processable vector data is available, to which the real database is associated (Figs. 12.9 and 12.10), element by element.

Finally, all the data have been arranged for the 3D visualization and query, always keeping the BDTRE structure (Figs. 12.11 and 12.12) and allowing a detailed representation of street furniture, trees, and so on.

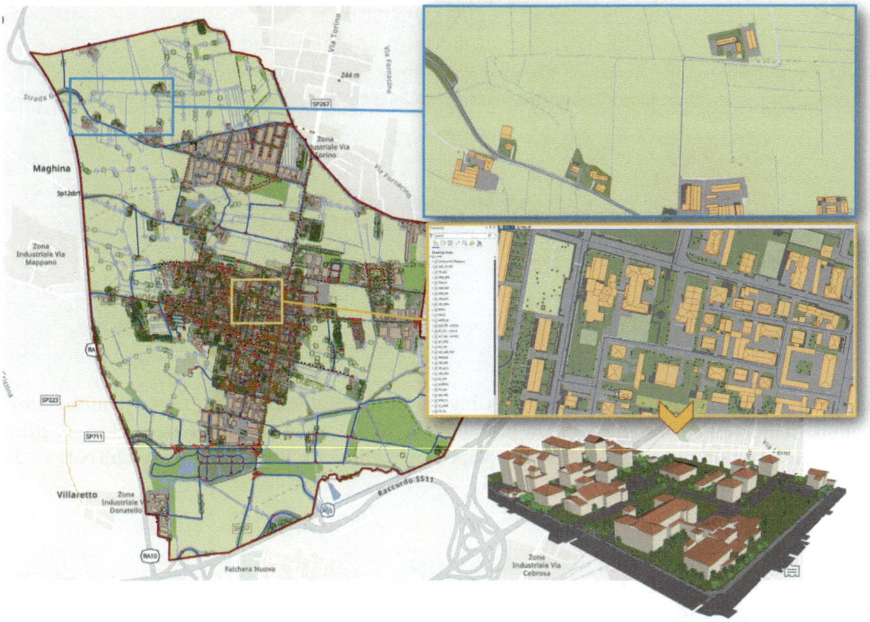

Fig. 12.8 GIS project of the Mappano municipality. ArcGISPro screenshot, taken from the GIS project developed by authors for the present research (both in 2D and 3D)

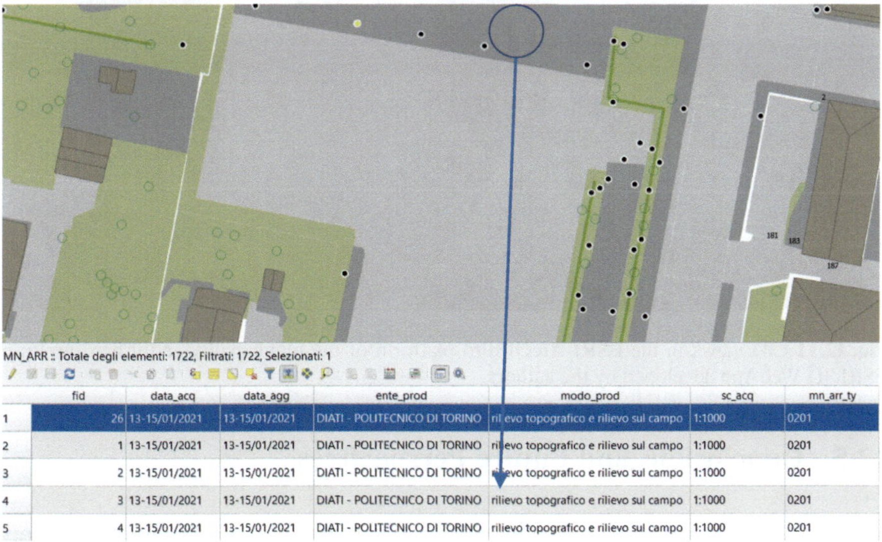

Fig. 12.9 Data query reports information of each geometry. ArcGISPro screenshot, taken from the GIS project developed by authors for the present research (both in 2D and 3D)

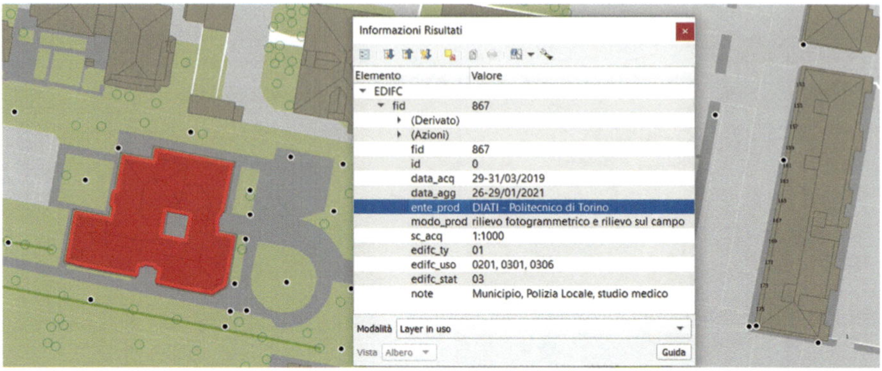

Fig. 12.10 Data query is possible also bypassing the attribute table using the "element information" function. In this case, one single window accounts for the values of one single element. ArcGISPro screenshot, taken from the GIS project developed by authors for the present research (both in 2D and 3D)

Fig. 12.11 3D views in the ESRI ArcGISPro platform of the municipality and urban furniture. ESRI 3D WebApp developed by the authors

12.5 Conclusions and Future Perspectives

The methodology here presented shows an innovative workflow. Thanks to the adoption of 3D metric survey techniques and data outputs, it is possible to help the definition of a high detailed and accurate digital cartography. The digital GIS twin shows the reality in 2D and 3D, representing all the city, land, and territory objects with their information stored into a standard-based geoDB.

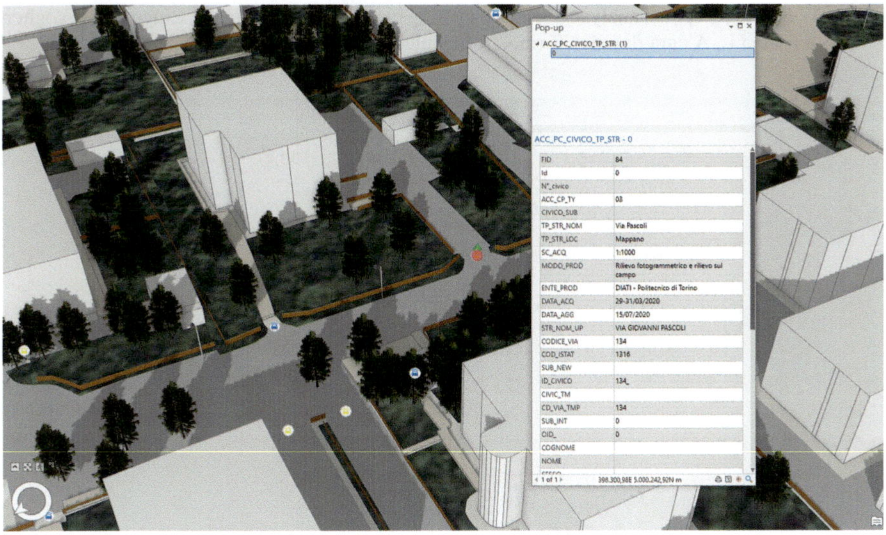

Fig. 12.12 3D GIS query of 3D objects with related information stored into the attribute table for the civic access. ESRI 3D WebApp developed by the authors

Future perspectives and works could include developing user-friendly tools and webGIS applications. It will be possible to query, measure, and analyze the spatial data of the municipality. Moreover, a WebApp map could be adopted by different stakeholders (such as citizens, technicians, public administrations) for various purposes (e.g., energy information of buildings, urban planning, etc.).

Acknowledgements The authors would like to thank Ing. Vincenzo Di Pietra for his support to the Master thesis of Ing. Egle Beani, the administration of the Municipality, and the Association of Volunteers "Arcobaleno Mappanese" that helped during the surveying activities.

References

Beani E (2019) Produzione di cartografia tecnica in ambito comunale mediante. Il caso del Comune di Mappano. (V. Di Pietra, A cura di) Torino, Politecnico di Torino

Biljecki F, Ledoux H, Stoter J (2016) An improved LOD specification for 3D building models. Comput Environ Urban Syst 59:25–37. https://doi.org/10.1016/j.compenvurbsys.2016.04.005

Geoportale Nazionale (2021) Geoportale Nazionale. Tratto da Normativa. http://www.pcn.minambiente.it/mattm/inspire-normativa-di-riferimento/

Gröger G, Kolbe TH, Nagel C, Häfele K-H (2012) OpenGIS city geography markup language (CityGML) encoding standard, Version 2.0.0. OGC Document No. 12–019, pp 344

Pileri P, Scalenghe R (2016) The impractical supremacy of local identity on the worthless soils of Mappano. City, Territory and Architect. https://doi.org/10.1186/s40410-016-0035-z

Shahat E, Hyun CT, Yeom C (2021) City digital twin potentials: a review and research agenda. Sustainability 13(6). Tratto da. https://doi.org/10.3390/su13063386

Traore A (2021) Geospatial analysis for urban cycling planning and design. In: Lingua AM, Voghera A, La Riccia L, A cura di (eds) Torino, Politecnico di Torino

Chapter 13
Source and Data for the Analysis of the Metropolitan Territory with GIS Tools: A Critical Review Between Commercial and Open Access Tools

Francesco Fiermonte, Luigi La Riccia, and Mattia Scalas

Abstract This contribution addresses the issue of the availability and accessibility of data for urban and regional studies. After an initial part in which tools and arguments are provided to frame the open data paradigm, the second part proceeds with the simulation of a GIS-based study on a territory of the Metropolitan City of Turin aimed at supporting post-COVID territorial planning, highlighting the areas in which there is a lack of data, the difficulty of access, such as in health care, or the obligation to have to use commercial data. In the last part, the results of the study are discussed, and the issue of the importance of the construction of digital models (digital twins) capable of receiving and processing data of different sources, and how these models should be developed according to an open and non-commercial paradigm.

Keywords GIS · Spatial planning · Geospatial data · Open data · Digital twins

13.1 Introduction

The analyses aimed at spatial planning for resilience, which pursue the objectives of sustainability and reduction of vulnerabilities, require more and more data of different origins, to be integrated to create models, scenarios and prefigurations useful to interpret complex and closely related phenomena (Voghera and La Riccia 2018).

F. Fiermonte · L. La Riccia · M. Scalas (✉)
Inter-university Department of Regional and Urban Studies and Planning, Politecnico di Torino, Turin, Italy
e-mail: mattia.scalas@polito.it

F. Fiermonte
e-mail: francesco.fiermonte@polito.it

L. La Riccia
e-mail: luigi.lariccia@polito.it

G. Brunetta et al. (eds.), *Post Un-Lock*, The Urban Book Series,
https://doi.org/10.1007/978-3-031-33894-6_13

The quality of analyses itself depends on the available data, and that's why the rise of open data, open source software and new types of licences seem to be an attempt to prevent the monopoly of data by a small number of subjects in order to promote the dissemination of data and the possibility of gaining knowledge.

However, the availability of truly open data sometimes seems to be limited and suffers from a lack of accessibility and systematicity that can facilitate user access and consultation. The objective of this contribution is to recompose the theoretical framework underlying the concepts of open data, open source and creative commons and the reasons why they can support planning, in order to provide the reader with a synthetic scheme useful to orient himself in the data search phase aimed at analysis for territorial planning, and to critically analyse whether the availability of open data is sufficient to allow in-depth and comprehensive analysis of a territory and within what margins it is necessary to resort to commercial or non-open data emphasizing the need for a paradigm shift towards the open model.

13.2 Background: Open Data in the Big Data Era and Its Relevance in Spatial Analysis

13.2.1 Data as a Common Good

> Everything is related to everything else, but near things are more related than distant things. (Tobler 1970)

Data are the basis of a large number of human activities, and are considered, not only in the collective consciousness but also at the legal level, increasingly a fundamental element for the performance of the same. We could define them as the oil of contemporary civilization, as they are an objective starting element for the construction of information, to be considered instead subjective as the result of subsequent elaborations on the "raw" data.

The use of data to derive information is fundamental in the public sector and in support of administration, in the private sector for the start-up and enhancement of economic activities and for the conscious participation of citizens in the public debate (Voghera and La Riccia 2018).

In this sense, access to and availability of data can be considered engines of stimulus and development for the economy, with particular reference to the service sector. In this regard, there is a tendency to consider the whole Internet as global public good.

Perhaps this is also the reason why a type of licence widely used in the digital world, the "Creative Commons"[1] set up by Lawrence Lessig and widely engaged in the promotion of the commons code, expressly refers to the "commons", those "common goods" that, in the field of natural resources, were the subject of the debate

[1] https://creativecommons.org/.

between Hardin (1968), who, speaking of "tragedy", argued that if individuals relied on themselves alone, and not on the relationship between society and man, then people will treat other people as resources, which would lead to the world population growing and for the process to continue; and Ostrom (1990), who instead highlighted their nature, neither public nor private, but collective and self-managed by the users themselves.

13.2.2 Volume of Data and Information Extraction: Big Data

The term data-driven is often used not only for companies but also for markets, indicating the propensity to rely on data to make decisions that are as objective as possible. To speak of data-driven entities, it is necessary to generate and dispose of large quantities of data, acquired in such a way as to guarantee their validity and adherence to objective reality. Regarding the first aspect, the amount of data available, it is easy to identify a strong correlation between their availability and the technical capacity to acquire and store them. In this sense, storage and computing capabilities are correlated in turn to the improvement of hardware performance, well described by the empirical assumption of Moore's first law (1965):

> The complexity of a microcircuit, measured for example by the number of transistors per chip, doubles every 18 months (and then quadruples every 3 years).

The increase in hardware performance, coupled with a drop in device prices, has led to an increase in the number of players who can create and manage increasingly large datasets, leading to an exponential growth in the amount of data produced annually throughout the planet.[2]

We can consider the increased availability of data and, consequently, of information as one of the most obvious manifestations of the technological revolution we are experiencing. In recent years, the term "big data" has thus become popular. Often misused, this expression generally indicates any collection of data so extensive in volume, velocity and variety that special technologies are needed to extract information from it (De Mauro et al. 2016). In a nutshell, a human operator without tools would not have the ability to extract information from this data.

Examples of big data are the logs of accesses to a website, the profiles of a social network, the list of transactions made by customers of a large online store and their personal information, the set of surveys made by a satellite. All datasets that would be unusable, or very slow to process, without the help of special algorithms that allow us to obtain useful information. The crossing of the data of the dataset or between different datasets, in turn, allows to automatically discover unexpected correlations, which are then validated or discarded by a human operator on the basis of the well-known principle "correlation does not imply causation".

[2] https://www.statista.com/statistics/1183457/iot-connected-devices-worldwide/.

The amount of data available is so large that most of those collected are never analysed, generating a real data gap. This amount of data is known as "dark data", a term coined by Gartner Inc.[3] to indicate data that is collected and processed but never used. The use of these amounts of data is one of the factors that led to the development of data mining. The term "data scientist" was first used by Dhanurjay "DJ" Patil, a computer scientist and Chief Data Scientist at the US Office of Science and Technology Policy, and indicates the process of extracting non-trivial, previously unknown and potentially useful information from available data. It is interesting to note, as evidence of what was written at the beginning about data as the oil of contemporary civilization, that the term mining seems almost to consider data an external entity as much as a mineral deposit generated naturally underground. This term almost seems to recognize that the ability to acquire data has far exceeded the ability to keep track of it and obtain information from it without a specific "exploration": a mining, in fact.

13.2.3 Where Is Big Data Coming From?

Who generates data? As we have seen above, data are acquired in a more or less systematic way by many actors, from public administrations to companies or non-profit organizations. Basically, any organization produces data.

To give some examples, every query made on a search engine generates data on who made this search. Similarly, the "path" of the pages visited during navigation generates a series of data that, if properly interpreted, allow those who own them to reconstruct the evolution of intentions and choices made by users: information that can then become precious to determine—for example—which advertisements to visualize in the spaces of the websites hosting the advertisements.

In concrete terms and using limited free tools, from a simple comparison between the words ArcGIS and QGIS elaborated through Google Trends,[4] we can observe the diachronic trend and the territorial areas most interested by the searches of the users of these two words and reconstruct the evolution of the interests and the possibility to generate value in terms of requested services.

Another example of big data relates to data acquired by satellites. As of April 2020, there are 2666 satellites in active orbit around our Planet, of which 1440 have commercial purposes and 436 are governmental. There are 339 for military use, 133 civilian and 138 mixed (Space Foundation 2021).

Each of these satellites has a specific purpose, ranging from the maintenance of GPS systems to telephony and direct observation of the Earth. With regard to the latter, it is sufficient to mention some of the most famous government satellites to realize the amount of data generated: the American Landsat, the European ERS and Sentinel acquire every day hundreds of megabytes in their orbits around the planet,

[3] https://www.gartner.com/en/information-technology/glossary/dark-data.

[4] https://trends.google.it/trends/?geo=it.

generating useful data to monitor the changes taking place on the surface. On a commercial level, we should mention the services offered by QuickBird, IKONOS and WorldView are able to offer customers high and very high resolution satellite images.

13.2.4 Data Availability and Access: Some Critical Issues

This data revolution (Voghera La Riccia 2018), however, raises some relevant issues that need to be exposed in order to fully understand the meaning of the whole "paradigm" of open access data and open source software. We could summarize these issues in four expressions: ownership, access, usability (or reusability) and privacy.

It seems quite logical to assume that the owner of a given data set is the one who generated it. A private service that, after having obtained the appropriate governmental authorizations with its satellite, acquires images has every right to use them as it sees fit and to eventually obtain a profit. However, the problem from this point of view is represented by the disparity in the possibility to generate and cross data. In this sense, we can use the case of Google as a useful example. The Californian Company is the owner of the most popular search engine in the world: in essence, it has the ability to process millions of queries from its users and cross-reference the information in order to obtain useful information. In 2005, the company launched the Google Maps service, a webgis that covers a large part of the Earth allowing to visualize geographical maps.

Over time, the service has expanded considerably, and today it allows to view not only the basic cartographic layers, but also to build itineraries and to identify stores and other economic activities added by those directly concerned through the Google Places service. The company, strengthened by its leadership position among search engines, has introduced for the user the possibility of evaluating and expressing his own opinion on the places visited, contributing to define the overall image that a given economic activity has for Google users and in fact conditioning the choice of possible future customers of that particular service. Crossing further this information with that made available by the Street View service, it is possible to make available to Google's customers, the merchants in this case, a series of cognitive tools useful to optimize the possibilities offered by the "economic positioning" of their business with access to information generated—for free—by the same Google users.

It is also true, however, that business intelligence, especially if directed at geomarketing, which also distils information from a "location analysis" is not shared for free but, like all of Google's advanced services, is subject to precise pricing for—in this case—the merchants involved.

The program was further strengthened in 2015 with the creation of the Google Local Guides service, which in exchange for some benefits from the company such as free cloud space for the user, allows the company to obtain additional information about the places visited. In fact, joining the program implies from the user the possibility to answer specific questions about the accessibility of the place, the level of crowding, the services offered.

This ability to acquire user-generated data for free, both unintentional (such as browsing histories) and voluntary (content posted on a social or ratings of a place) puts large tech companies at a significant advantage over potential new competitors and sometimes over governments themselves. Access to the information generated by this data, in fact, is in fact access to proprietary information, given to the provider company for a financial consideration or in exchange for data, information, or a licence to use it (such as photos posted on Instagram).

This modus operandi has generated considerable debate around the issue of privacy, prompting governments and supranational institutions to legislate on the subject. The most important example in Europe is in this regard the General Data Protection Regulation 2016/679 (GDPR),[5] published in the Official Journal of the European Union on May 4, 2016, and in implementation since May 25, 2018. The regulation harmonizes national regulations, and shifts the focus from a proprietary view of data to one based on user control of the data, encouraging free movement and the right to know the nature and use of personal information in the hands of third parties. The regulation applies to both automatically generated and non-automatically generated data and effectively requires the use of an informed consent form prior to the granting of one's data.

While the GDPR has attempted to remedy some asymmetries in data generation and transfer—for example, users now have the option to avoid tracking their browsing data for marketing purposes by denying consent to a website's use of cookies—on the other hand, this regulation poses generalized difficulties in the acquisition and cross-referencing of data by even non-profit entities such as universities and research institutes and does not solve the problem of data access.

In a society where the use of data is increasingly fundamental to the development of activities and knowledge of reality, this issue is crucial. The ownership of information and the possibility of accessing it are—and will be even more so—a necessity for those who wish to start or maintain an economic activity, or even more so for study and research purposes, perhaps aimed at supporting administrative action.

If data is therefore seen as something necessary for the development—in fact—of society, it is necessary to remedy the inhomogeneity in distribution and access. Initiatives aimed at improving this aspect are often characterized by the use of the "paradigm" of open data. The Open Knowledge Foundation[6] has elaborated the open definition:

> Open data and content can be freely used, modified, and shared by anyone for any purpose.[7]

---

[5] https://eur-lex.europa.eu/EN/legal-content/summary/general-data-protection-regulation-gdpr.html.

[6] https://okfn.org/.

[7] https://opendefinition.org/.

The availability of freely accessible data is based on the idea of increasing transparency, releasing social and economic value and increasing community participation and engagement. Among the most useful data in this sense are of course spatial data, i.e. any data that can be associated with information in space. Rather than to ensure open access to spatial data as much as to promote the sharing of a common infrastructure, the European Union has issued the INSPIRE Directive,[8] which requires member states to "systematize" interchangeable geographic data, metadata and services to facilitate access and reuse. These data are usually derived from satellite acquisitions, performed as seen above via the Copernicus system[9] and Sentinel satellites, subsequently reprocessed by the European Space Agency (ESA) and other institutions.

In parallel, many volunteers and non-profit organizations have been engaged in building datasets "from below", according to a participatory and "active citizenship" paradigm. The most important example is in this sense the OpenStreetMap project.[10]

The idea of open access is not limited only to the data itself, but also concerns the processing capacity. We therefore speak of open source, meaning all free software and algorithms whose source code is freely accessible and modifiable. It goes without saying, in fact, that without the ability to process information without cost (or at least without significant costs), access to the data itself would become almost useless. The community that has built and maintains the QGIS spatial data processing software is committed to this. In this regard, it is worth mentioning the community of volunteers who implement programming languages, a fundamental element both for the development of software and for the performance of more advanced operations of reading and analysis: in this regard, we cite the vast community of R, Python, PostGreSQL and the countless "packages" of code that, when installed, allow you to perform special operations on your GIS software, such as WhiteBox Tools. The "open" community is growing and has greatly influenced the way geospatial data are acquired, processed, analysed and visualized (Coetzee et al. 2020).

This set of factors, the availability of public data and the large community of volunteers who carry out "open" projects, together with the improvement of computational performance at lower costs, has greatly expanded the ability to analyse the territory not only by experts and academics but by anyone who is "familiar" with these computer tools. At this point it is mandatory or, more simply, likely to incur costs if you choose to use proprietary software, to use an external technical support or need a computational power not available, thus having to resort through cloud computing services to external service providers.

The question now is whether the available data alone are sufficient to meet the needs of increasingly accurate analysis and studies of the territory, fundamental in planning if we decide to adopt a contemporary paradigm aimed at developing the resilience of a territory. Are we really at that point? Are we at an optimal level that

[8] https://inspire.ec.europa.eu/inspire-directive/2.

[9] https://www.copernicus.eu/en.

[10] https://www.openstreetmap.org/.

can be improved, or are we even at a guard level below which analyses risk losing quality, forcing us to rely on proprietary platforms, data and services?

To be able to work in this sense, in fact, or rather to ensure that territories are planned in such a way as to be aware of their vulnerabilities, ready to deal with risk and adapt to change, it is essential to have an increasingly in-depth knowledge of phenomena at various scales, their relationships and how they change over time. It is therefore necessary to be able to build models of the territory, where multiple information of various derivations can be crossed to appreciate the complexity and the network of relationships. The ability to account for multiple elements in parallel is, for example, a feature of modern multivariate analysis, but this methodology needs to rest on robust, large and reliable data systems.

13.3 Availability and Access to GIS-Based Data on the Metropolitan City of Turin

One of the main tasks of spatial analysis for planning is to provide data and information on the territory under study and planning. Data retrieval and the construction of information and interpretation plays a fundamental role in the process of interpreting the territory itself, which allows the implementation of robust and place-based strategic or planning tools.

If the intention is to adopt a territorial resilience approach, it is also necessary to use a multidisciplinary approach that responds to the technical challenge for building new knowledge tools that can support the development of strategies, plans and actions aimed at reducing local vulnerabilities (Beltramino et al. 2022). The arrival of the COVID-19 pandemic has also opened a window of reflection on what data can be further sought.

Trying to respond to these stimuli and trying to prioritize free access to data, we have hypothesized a study of the metropolitan area north of Turin by searching among the databases available data that were useful in order to provide a valid analytical support to the objective.

13.3.1 Satellite Images

A first element to check availability is related to satellite images in raster format. As pointed out in the first part of this paper, many public bodies provide good quality images useful for carrying out a series of supporting analyses. In particular, it is possible to download images from the Copernicus portal from Sentinel satellites, USGS and European Space Imaging search engines. Other satellite images, available for a fee, are made available by the Italian Military Geographical Institute (IGM) and commercial companies. The usefulness of satellite photo analysis can be found for the calculation of some important indicators on soil consumption, soil slopes or physical phenomena (erosion, landslides, etc.).

13.3.2 Physical and Basic Cartography

We could define the basic cartography as the set of elements necessary to provide an immediate "eye" on a given territory: type of land, level curves, buildings, roads and other elements. As far as Piemonte is concerned, the data source is represented by BDTRE, distributed under Creative Commons—BY 2.5 licence, in progressive replacement with version 4.0. Built in implementation of the 2007 INSPIRE Directive, with the Regional Law of 5 February 2014, it has become the reference cartographic base for all public and private entities that interface with the body itself. BDTRE offers numerous information layers in the traditional shapefile format and in the new geodatabase.

Another unofficial data source is the OpenStreetMap project. Other basic cartography can be found through the national geoportal and other sites of research centres and governmental institutes.

13.3.3 Demographic Data

The analysis of demographic data is usually carried out in Italy on the basis of data provided by the censuses of the Italian National Institute of Statistics (ISTAT),[11] freely accessible and downloadable from the portal of the Institute under licence CC BY 3.0 IT: at the moment, data sets of territorial bases and census variables of population censuses and industry and services are available. The censuses available in this way are those of 2011, 2001 and 1991 and the minimum territorial base is represented by the census section. It has been noted that in some cases, it is difficult to extract information from ISTAT demographic data due to inconsistencies between the spatial bases in the form of shapefiles and those tabular with the socio-economic information, and it is noted that the ten-year period of the censuses can make the data old compared to the type of analysis for which they are needed. In this sense, the entry into operation of the new permanent census could be of great importance in order to improve the quality and quantity of studies carried out.

[11] https://www.istat.it/.

13.3.4 Vehicular Flows

Data on vehicle flows have been made possible for years by direct measurements at the entrance and exit of a given road trunk. With the spread of the "black boxes", the GPS devices used to reconstruct the dynamics of possible accidents and on which to calibrate the rates of car insurance and other tracking devices the monitoring of this phenomenon has become easier, but the data is not immediately accessible.

In Piedmont, this kind of information is in the possession of 5 T,[12] an in-house company with an entirely public participation.

13.3.5 Soft Mobility

Researching data on gentle mobility, understood as the set of journeys made on foot, by bicycle or by other physical traction vehicles, can be of great interest as much as it is difficult to succeed. First of all, how do you measure these movements? As for those on foot or by bicycle, the answer comes from the world of sport, and from the widespread spread of Smartwatch, Running Watch, Cyclocomputer and similar or directly from popular smartphones. These devices have been accompanied by applications for several years and after registration by the user provide both a tracking of their movements free of charge—provided by the wearable device and then downloaded to the app—or directly recorded by the app installed on the phone, both statistics on the type of activities such as metres travelled, average speed, an estimate of the watts produced and so on. It is also possible for you to share this information on specialized social networks, such as Strava.[13]

The set of user data, as easily imagined, goes to build huge datasets of aggregated information that allows to obtain information about lifestyles, habits, movements. In order to support spatial planning, access to these data sets would be of particular importance in order to obtain information on, for example, the use of green areas, the use of cycle paths or the finding of routes and routes that exist de facto but have not been established by the public authority.

These data sets are privately owned, but there are some services that allow them to be accessed by professionals and researchers. In particular, there are two Strava instruments, the Strava Global Heatmap and Strava Metro. Strava Global Heatmap is a webgis that represents the "travel density" of a given path by users who upload their activities to the social network. In extreme synthesis, it consists of an OpenStreetMap cartographic base, which is gradually replacing Google Maps, and with a "style" of terrain rendered through MapBox, on which appear in aggregate form the activities of users, thematized by intensity of colour. The more intense the colour, the more that stretch was crossed by Strava users. Strava's Heatmap is used by OpenStreetMap users to improve their service and is available in JOSM by loading a TMS level.

---

[12] https://www.5t.torino.it/.

[13] https://www.strava.com/.

However, the Heatmap does not allow any other analysis, even with the help of colour-based raster analysis tools, so the use of Strava Metro is necessary. Metro is a service born in 2016 as a support tool for city planners and other experts, and provides data sets that can be used for the study of flows, road safety and project evaluation. However, access is subject to a request from the company, with long times and undiscounted outcomes.

Other systems for estimating pedestrian flows are based on the estimation of the number of movements in relation to the areas of attendance (green areas, administrative offices, etc.).

13.3.6 Energy Consumption

The analysis of energy consumption can be considered an important indicator useful to outline strategies for the territory and the city such as refitting, plant modernization and strengthening self-production. A rapid research on Piedmont on the website of the Agenzia per l'Italia Digitale[14] shows little data, mainly related to the electricity consumption of school buildings in the Metropolitan City of Turin and the consumption of gas on a municipal basis.

The consumption data relating to users are the property of the companies that manage the contracts, such as IREN[15] in Turin. The data, once requested and obtained, are provided in the form of.csv with annual and monthly consumption. The rows are not georeferenced but associated with an address and require geocoding operations to obtain a precise shapefile that allows them to be used in GIS environment.

13.3.7 Risk and Hazards

A large number of public and open access datasets are available for risk and hazard analysis. For example, it is possible to estimate the fire risk by calculating the IPSI indicator that crosses land cover data (e.g. vegetation types), the slope of the land and the construction characteristics of buildings. If an on-site investigation may be required on the last point, it is also true that the remaining data—DTM, soil cover maps such as Corine Land Cover—are easily accessible. The flood hazard can be estimated with the ranges made available by Hydrogeological Asset Plans, as well as the danger posed by landslides. Other interesting data sets can be found on portals such as ARPA[16] for air quality monitoring.

---

[14] https://www.agid.gov.it/.

[15] https://www.gruppoiren.it/.

[16] https://www.arpa.piemonte.it/.

13.3.8 Health Data

The search for health information is particularly difficult, with data that is in public hands but difficult to access. A practical example is access to data related to the spread of COVID-19 infections. Available and publicly accessible data can be accessed on a municipal scale, making any kind of analysis complex at a more detailed scale. The problem in this case is mainly related to privacy and to the set of rules that protect the individual.

13.3.9 Presence of Services

Defining the word "services" is complex, but in general they are quite simple to find through the regional geoportal all those elements related to the public: schools, parks, but also churches, sports facilities, monuments, hospitals and sanitary facilities. These data can normally be identified within BDTRE, or with the discharge of other shapefiles from regional geoportals or metropolitan level ones.

However, if we use the term "services" in a broader sense, including, for example, commercial activities and merchants or non-profit organizations, the theme becomes more complex.

In this case, the resources made available by open data and commercial projects are supported. The aforementioned OpenStreetMap is able to provide data—built by users—on commercial activities, classifying them according to the type of service offered. If you work in QGIS environment, it can be of great help to instal the OSM plugin that with a simple query allows you to download entire levels of open data for the territory you are studying.

Among the commercial services it is worth mentioning Google Maps, which as already mentioned above is also able to provide user-generated information to correlate searches on its search engine. Maps information can be accessed for a fee via the Google Maps Platform service from Google Cloud.

13.4 Conclusions

The development of a comprehensive city-wide model, which integrates data from various sources, to provide valuable information supporting initial planning does not necessarily require the use of proprietary or restricted data.

The integration of basic, flow and risk information into a GIS database would make it possible to build an effective decision support system and should be seen as the arrival point for the development of new analysis methodologies.

However, the possibility of accessing the data should be further implemented, making it easier to consult and download data of public origin at least for research or support to public administrations. The sectorization characteristic of many sectors of public administration therefore also affects the integration of data, thus conflicting with the very aims of planning that seeks to operate according to multilevel and multi-sector logic.

Integration in this regard should be pursued by strengthening the infrastructure that seeks to link datasets of different origins, ensuring its publicity and the widespread adoption of the open access paradigm with rights allocation that allows the re-use of data also for commercial purposes. The dissemination of integrated public infrastructures, which provide "free" data, could be an element of interest not only for the study and monitoring of the territory but also for the development of new economic opportunities, bypassing in a certain sense the incoming costs represented by the need to use data but to be able to access it only for a fee.

From an almost "futuristic" perspective, the dissemination of real Digital Twins, digital city models that can provide updated and freely accessible data on multiple levels of authorisation would be desirable. Thinking from this perspective could be the only way to avoid the risk of data monopoly by other actors, with consequences for example on competition. In the background, a serious issue related to this model is related to privacy, and it will increasingly be necessary to develop new models capable of reaching a reasonable compromise between the need to acquire and access data and the protection of the individual. In this sense, a strong public role in the process of data acquisition itself could be an element of greater stability and reliability.

References

Beltramino S, Scalas M, Castro Rodriguez DJ, Brunetta G, Pellerey F, Demichela M, Voghera A, Longhi A, Mutani G, Caldarice O, Miraglia G, Lenticchia E, La Riccia L (2022) Assessing territorial vulnerability. Testing a multidisciplinary tool in Moncalieri, Italy. Tema 15(3):355–375

Coetzee S, Ivánová I, Mitasova H, Brovelli MA (2020) Open geospatial software and data: a review of the current state and a perspective into the future. ISPRS in J Geo-Inf 9:90

De Mauro A, Greco M, Grimaldi M (2016) A formal definition of big data based on its essential features. Libr Rev 65(3):122–135

Garrett H (1968) The tragedy of the commons. Science 162(3859):1243–1248

La Riccia L, Ardissono, Voghera A, Semantic spatial representation and collaborative mapping in urban and regional planning: The OnToMap community project. In: Voghera A, La Riccia L (eds) Spatial planning in the big data revolution, IGI Global, Hersh (USA), pp 194–220. https://doi.org/10.4018/978-1-5225-7927-4.ch009

Moore GE (1965) Cramming more components onto integrated circuits. Intel Electronics Magazine. Available via https://web.archive.org/web/20190327213847/https:/newsroom.intel.com/wp-content/uploads/sites/11/2018/05/moores-law-electronics.pdf (Accessed 16 Jan 2022)

Ostrom E (1990) In: Governing the commons: the evolution of institutions for collective action. Cambridge University Press, pp 1–9

Tobler W (1970) A computer movie simulating urban growth in the Detroit region. Econ Geogr 46(Supplement):234–240

Voghera A, La Riccia L (2018) Sharing semantic territorial interpretation: the contribution of OnToMap project to spatial planning. New approaches, methods, and tools in urban E-planning. Advances in civil and industrial engineering book Series, Springer

Part V
Lesson Learned and Perspectives

Chapter 14
Final Remarks on the Implementation of the Post-pandemic City and the Role of Technology

Patrizia Lombardi

Abstract This final chapter of the book highlights some major results of the Post Un-Lock research project conducted inside the SDG11Lab and the Responsible Risk Resilience Center of the Politecnico di Torino with the collaboration of the Medical Statistics and Epidemiology Department of the University of Turin. In particular, it discusses the concepts of proximity and Local Resilience Units in connection with the well-known "15-city" paradigm. This model requires the adoption of new urban and ecosystem criteria as well as advanced technological support systems for post-pandemic public and private urban space planning and management. The conclusions point out to the need for education and appropriate digital skills in order to be able to integrate ecological principles in the design and management of urban areas with the support of high-tech systems.

Keywords Proximity · SDGs · Technology · Innovation · Competences

This book has provided some major results from the Post Un-Lock research project conducted inside the SDG11Lab and the Responsible Risk Resilience Center (R3C) of the Politecnico di Torino with the collaboration of the Medical Statistics and Epidemiology Department of the University of Turin. The research has highlighted the importance of increasing the capacity of cities and communities to adequately respond to crisis and catastrophic events like the COVID-19 pandemic by enhancing co-evolutionary resilience and integrating ecological principles in the design and management of urban areas. In particular, the research has specifically pointed out the need for urban and regional developers and decision-makers to adopt new urban and ecosystem criteria and parameters for post-pandemic public and private space planning and design (Mirabella et al. 2019). Furthermore, the concept of Local Resilience Units (URLs), defined as the minimum system capable of responding to the effects of crises, has been pointed out as an opportunity for improving the quality of life of

P. Lombardi (✉)
DIST—Interuniversity Department of Regional and Urban Studies and Planning, Politecnico di Torino, Turin, Italy
e-mail: patrizia.lombardi@polito.it

G. Brunetta et al. (eds.), *Post Un-Lock*, The Urban Book Series,
https://doi.org/10.1007/978-3-031-33894-6_14

residents by promoting social life and reducing consumption of resources and energy. Local Resilience Units are based on the concepts of urban accessibility, proximity and walkability, representing important features of contemporary cities, which are required to adopt sustainable development models, in line with the Agenda 2030 objectives and specifically with the Sustainable Development Goal (SDG) 11 on Sustainable Cities and Communities. Table 14.1 provides the specific 7 targets and indicators used for monitoring the implementation (UN 2022).

By reducing the need to long-distance travel and ensuring fairer access at the local level to urban services for a wide range of citizens, URLs may be able to help the achieving of a number of the above targets.

As pointed out by Pede et al. (2023), the concept of URL is connected to the so-called 15-min-city model. This refers to a transition toward more livable, resilient and inclusive cities, characterized by higher and equally distributed levels of accessibility to services (Moreno et al. 2021). It aims to reduce dependence on the car both in terms of use in favor of other modes of transport (modal shift) and in terms of ownership (car ownership), working on the two dimensions of time and space, i.e., on an optimal distribution of services in space and planning the time needed to reach them in a sustainable manner. It presupposes a "hierarchization" of services at the neighborhood and urban scale distinguishing between: "neighborhood services," e.g., retail, primary secondary education, … and "basic services," e.g., hospitals, universities, etc.

Neighborhood services should be reached on foot or by bicycle in 15 min, while basic services should be located in areas accessible within 15–30 min by using rapid mass transport (metros, trams, rapid transit buses, …), and where this is not sustainable (e.g., economically) by promoting forms of shared mobility also by car (car-sharing, taxis, dial-a-ride services, …). Therefore, this model should be interpreted as an approach to integrated transport-land-use planning, which aims at a reorganization of public spaces and services in favor of active mobility and micro-mobility, and at the same time reduce car dependency by supporting forms of collective and shared mobility.

Widespread proximity and accessibility to services through active, collective and shared forms of mobility has a direct impact also on Goal 10 (reduced inequalities) by helping to ensure equal access to work, education, health care, recreation and other social practices. On the other hand, it has an indirect impact on the goals concerning: health and well-being (SDG3), because walking and cycling are good for people's health and improve air quality; acting for the climate (SDG13), because promoting active, collective and shared forms of mobility reduces car travel and thus the emission into the atmosphere of dust and climate-altering gases such as carbon dioxide (note: car travel—in Italy—contributes to 24% of the total emissions of Co2 equivalent into the atmosphere).

As highlighted in the guidelines for National Urban Agenda developed by the Sustainable Infrastructure and Mobility Ministry (MIMS 2022), the realization of the 15-min-city paradigm requires technologies support. Technologies are key tools for the development of Cities of Tomorrow (Vandecasteele 2019) as these enable to: (i) acquire data from the field (all types of sensors); (ii) transport them bidirectionally,

Table 14.1 SDG 11 Goals, targets and indicators

Goals and target	Indicators
11.1 By 2030, ensure access for all to adequate, safe and affordable housing and basic services and upgrade slums	11.1.1 Proportion of urban population living in slums, informal settlements or inadequate housing
11.2 By 2030, provide access to safe, affordable, accessible and sustainable transport systems for all, improving road safety, notably by expanding public transport, with special attention to the needs of those in vulnerable situations, women, children, persons with disabilities and older persons	11.2.1 Proportion of population that has convenient access to public transport, by sex, age and persons with disabilities
11.3 By 2030, enhance inclusive and sustainable urbanization and capacity for participatory, integrated and sustainable human settlement planning and management in all countries	11.3.1 Ratio of land consumption rate to population growth rate 11.3.2 Proportion of cities with a direct participation structure of civil society in urban planning and management that operate regularly and democratically
11.4 Strengthen efforts to protect and safeguard the world's cultural and natural heritage	11.4.1 Total expenditure (public and private) per capita spent on the preservation, protection and conservation of all cultural and natural heritage, by type of heritage (cultural, natural, mixed and World Heritage Centre designation), level of government (national, regional and local/municipal), type of expenditure (operating expenditure/investment) and type of private funding (donations in kind, private non-profit sector and sponsorship)
11.5 By 2030, significantly reduce the number of deaths and the number of people affected and substantially decrease the direct economic losses relative to global gross domestic product caused by disasters, including water-related disasters, with a focus on protecting the poor and people in vulnerable situations	11.5.1 Number of deaths, missing persons and directly affected persons attributed to disasters per 100,000 population 11.5.2 Direct economic loss in relation to global GDP, damage to critical infrastructure and number of disruptions to basic services, attributed to disasters 11.5.3 (a) Damage to critical infrastructure and (b) number of disruptions to basic services, attributed to disasters

(continued)

Table 14.1 (continued)

Goals and target	Indicators
11.6 By 2030, reduce the adverse per capita environmental impact of cities, including by paying special attention to air quality and municipal and other waste management	11.6.1 Proportion of urban solid waste regularly collected and with adequate final discharge out of total urban solid waste generated, by cities 11.6.2 Annual mean levels of fine particulate matter (e.g., PM2.5 and PM10) in cities (weighted population)
11.7 By 2030, provide universal access to safe, inclusive and accessible, green and public spaces, in particular for women and children, older persons and persons with disabilities	11.7.1 Average share of the built-up area of cities that is open space for public use for all, by sex, age and people with disabilities 11.7.2 Proportion of persons victim of physical or sexual harassment, by sex, age, disability status and place of occurrence, in the previous 12 months

Source https://sdgs.un.org/goals/goal11

i.e., to and from the user (including 5G and 6G systems); (iii) process the data, up to and including producing simulations, what-if analyses and real predictions (storage systems, processing at the nodes—edge—and at the center, data mining systems, predictive modelling, machine learning, advanced analytics); (iv) deliver services and information (applications and user devices).

With reference to this paradigm, urban planners, local authorities and decision-makers should significantly increase the equipped pedestrian areas and footpaths as well as develop and/or extend all the following: widespread bicycle lanes; rapid mass transport networks for connections to/from main urban poles; bike-sharing and electric recharging infrastructure at the neighborhood scale; modal interchange car parks at stations or other multimodal hubs; traffic-restricted zones, in which active mobility and shared mobility services using electric vehicles (cars, bicycles, other) can be promoted. Furthermore, the active participation of citizens and stakeholders in the design and implementation of mobility services and the use of public spaces (Mobility as a Service, MaaS 3rd level) should be implemented.

Understanding mobility needs through data collection and simulation of impacts is crucial as well as deepening knowledge of demand, including unexpressed demand, tools and applications that provide information to citizens and enable feedback on travel and needs.

The realization of the post-pandemic city requires availability of real time data and technology networks and infrastructures, cyber-physics resilience, innovation and multimodal approaches. It requires innovation procurement systems, public–private partnerships, data compliance, but also education and digital skills and competences. Education and training are key factors in fostering innovation in cities and supporting green and digital transition, inclusion and social impact. Without appropriate skilled and educated human capital, the probability to deliver a sustainable future is equal to zero.

References

Mirabella N, Allacker K, Sala S (2019) Current trends and limitations of life cycle assessment applied to the urban scale: critical analysis and review of selected literature. Int J Life Cycle Assess 24:1174–1193. https://doi.org/10.1007/s11367-018-1467-3

MIMS (2022) Ministero delle Infrastrutture e della Mobilità Sostenibili, 2022, Agenda Urbana del Ministero delle Infrastrutture e della Mobilità Sostenibili, Ottobre 2022

Moreno C, Allam Z, Chabaud D, Gall C, Pratlong F (2021) Introducing the "15-Minute City": sustainability, resilience and place identity in future post-pandemic cities. Smart Cities 4:93–111. https://doi.org/10.3390/smartcities4010006

Pede E, Scalas M, Staricco L (2023) Towards neighbourhoods as minimum units of resilience?, In: Brunetta, Lombardi, Voghera (eds) "Post Un-Lock. From territorial vulnerabilities to local resilience"

United Nations (2022) SDGs progress report 2022, Department of Economic and Social Affairs. https://unstats.un.org/sdgs/report/2022/Goal-11/

Vandecasteele I, Baranzelli C, Siragusa A, Aurambout J (ed) (2019) The future of cities EUR 29752 EN, Publications Office of the European Union, Luxembourg. https://doi.org/10.2760/375209 (online), https://doi.org/10.2760/364135 (print), JRC116711

GPSR Compliance

The European Union's (EU) General Product Safety Regulation (GPSR) is a set of rules that requires consumer products to be safe and our obligations to ensure this.

If you have any concerns about our products, you can contact us on ProductSafety@springernature.com

In case Publisher is established outside the EU, the EU authorized representative is:

Springer Nature Customer Service Center GmbH
Europaplatz 3
69115 Heidelberg, Germany

Batch number: 10844706

Printed by Printforce, the Netherlands